SCIENCE BETWEEN MYTH AND HISTORY

José G. Perillán received a joint doctorate in Physics and History from the University of Rochester in 2011. Currently, he is the Pauline Newman '47 Director of Science, Technology, and Society at Vassar College (Poughkeepsie, NY). He is an Associate Professor holding a joint appointment in the Physics and Astronomy Department and the Multidisciplinary Program on Science, Technology, and Society (STS). The present book on myth-history emerges from the author's experience teaching undergraduate courses in physics, history of science, and STS. In preparing his varied courses, Perillán moves fluidly among frameworks and modalities. He has harnessed the adaptability and empathy necessary for this relentless pivoting to carve out a common ground from which to reflexively engage the scientific past.

—

SCIENCE BETWEEN MYTH AND HISTORY

The Quest for Common Ground and Its
Importance for Scientific Practice

José G. Perillán

Vassar College, State of New York

With a Foreword by
Trevor Pinch

OXFORD
UNIVERSITY PRESS

OXFORD
UNIVERSITY PRESS

Great Clarendon Street, Oxford, OX2 6DP,
United Kingdom

Oxford University Press is a department of the University of Oxford.
It furthers the University's objective of excellence in research, scholarship,
and education by publishing worldwide. Oxford is a registered trade mark of
Oxford University Press in the UK and in certain other countries

Published in the United States of America by Oxford University Press
198 Madison Avenue, New York, NY 10016, United States of America

British Library Cataloguing in Publication Data
Data available

Library of Congress Cataloging in Publication Data
Data available

ISBN 978-0-19-886496-7 (Hbk.)
ISBN 978-0-19-888360-9 (Pbk.)

For my nucleus — Rebecca, Quino, and Eleni.

Foreword
By Trevor Pinch

Science is in trouble. This is not news. But what is news is that the trouble runs deep and that scientists have partly brought it upon themselves. With this timely book, an elegant series of case studies of controversies in modern science that returns science studies to some of its core concerns, the source of the trouble is firmly located. It lies in the myths science has inevitably created.

Science of course has delivered: whether seemingly miraculous cures to diseases or weapons to annihilate the world. And it has over and over again been shown to be grounded in human culture and history: it has its fads, foibles and fallacies; its power elites; its dalliance with eugenics; its ability to crush a Rosalind Franklin and to elevate to genius level, a Richard Feynman. It may profess to follow Robert Merton's norm of universalism but we know it is still all too often run by cabals of mainly white men. It is often no better than other professions on ethical grounds. From Kuhn to Oreskes scholars in history of science and now the new field of "science studies" have shown that science is less special than we once imagined.

But science is different in one crucial respect—it gets to write its own history. Sure bankers can write their own history but they don't need to as part of the process of lending you money. Science deals in something bankers cannot deliver: truth. Truth is not an exchangeable commodity—once you have it everything else becomes worthless or merely a primrose path. Scientists, as Thomas Kuhn first stressed, tell their history from the present viewpoint of the truth in science. Everything progresses towards today's truth. This is known as whig history. Aristotelian physics is not judged by the standards of its time but from what we know today. Slaying the dragon of error is the stuff of myth. José Perillán calls this process the writing of "myth-history".

But it is not just the content of the science which is mythologized—the process of getting there becomes part of the myth.

The story is well known. All it needs are a few heroic experiments; a brilliant breakthrough theory; a genius here a genius there; and all enshrined in Nobel prizes. A genius may need to stand on the shoulders of giants, as Newton proclaimed, but they are giants. The process is presented in a way such that human interests, in short anything to do with the stamp of humans, are merely accessories to the inevitable emergence of the truth. Why does science need to be anesthetized in this way? What is the purpose of what we in science studies call "boundary work"? It is because in a deep way in this telling of science, humans are necessarily effaced. As Peter Medawar showed years ago, it begins with the scientific paper and its peculiar literary construction. Rather than say what actually happened, science is cast in terms of disinterested language. Students are taught to write "observations were made" rather than a real human at a real time actually observed with her own eyes what was going on. In scientific controversies as Nigel Gilbert and Michael Mulkay showed a long time ago the truth is held to emerge, independent of humans, or as scientists will tell you: "the truth will out". To account for error all you have to do is point to the extraneous human intervention. The literary conventions surrounding science are all part of the myth. One of the surefire ways to discredit a scientist is to point to illicit human desires governing their work, whether seeking Nobel prizes, patent riches, or daring to adhere to a viewpoint after the rest of science has moved on. Humor will often help do the job. As the late Martin Gardner, who was a master at debunking pseudoscience, once said, "a joke is worth a thousand syllogisms" It is also part of how elite male physicists are trained to present themselves, as Sharon Traweek showed in her ethnography: think Jewish and act British but not the other way around.

Scientists themselves are of course aware of the myths. The greatest genius of them all, Albert Einstein, made the distinction between existing science, "the most objective thing known to man" and "science in the making" where all human foibles are exposed. Einstein himself, as recent historical accounts have shown, was far from the math-illiterate genius who could conjure up discoveries by pure thought alone. Scientists know that when professional historians tell the histories of their field it often aint pretty. But once the science is settled and dusted off everything falls neatly into place and the winners and losers can be adjudicated. Sure the winner may have been a little unsavory—recall the story of The Double Helix and James Watson the pushy yank at Cambridge pulling the wool over the eyes of the over polite Brits; his disparaging of the "blue stocking" Rosalind Franklin, his sneaking off with her X-ray diffraction results. But the Double Helix was the truth waiting to be discovered—nothing can stop Watson from being the hero of this narrative. And as Perillán notes, scientists need and feed off these discovery myths. Ironically one of the woman scientists he writes about, Jennifer Doudna, was a victim of myth history in an outrageous retelling of the discovery of CRISPR by Eric Lander, but was herself inspired to enter the field of biology by reading Watson's mythical account in The Double Helix.

Myth history, despite its appeal, does serious damage to science because it prevents us seeing science as it really is, warts and all. If we invest too much in it, we will inevitably be disappointed, especially as the public in the global pandemic comes up close to science perhaps for the first time. Like any good stories, myth histories enshrine power. The stories as Perillán nicely puts it: "curate and amplify a preferred historical signal. They filter out and further marginalize people and ideas that don't align with the status quo" (p. 69). The question the book sets out to answer is whether these powerful stories protect science and its social capital, or whether they in fact erode trust in science.

The dilemmas are evident in his case study of the fate of physicist, Joseph Weber, and the detection of gravity waves. It's a story of unbelievable scientific triumph. In 2016 physicists announced to the world the detection of a movement on the order of one ten-thousandth the diameter of a proton—the remaining ripple produced from two black holes colliding 1.3 billion years ago. The story is mythologized by evoking Einstein as the solitary genius, the giant shoulders upon which today's physicists stand, and invoking a back-of-the-envelope calculation where Einstein supposedly predicted such waves existed (other scientists had predicted them earlier and Einstein didn't believe the waves he predicted were detectable). But the triumphalist narrative takes a darker turn with the story of Weber.

A maverick second-world war veteran trained in microwaves, Weber just missed out on the discovery of masers and lasers. But, after inspiration from the physicist John Wheeler, Weber showed that gravity waves might be detectable on Earth and, in the 1960s, developed a program at the University of Maryland constructing what were in effect two huge aluminum bar "tuning forks", which would "ring" together when a gravity wave resonated through them. These massive bar detectors at two locations in America were shielded from terrestrial noise. With Weber reporting positive results, other groups rushed to build different sorts of gravity wave detectors. By and large they failed to find the signals Weber claimed (the full story is told in the book Perillán draws upon by sociologist, Harry Collins, Gravity's Shadow). Weber kept improving his techniques and apparatus but, as the consensus turned against him, he slowly lost support. He was ostracized and eventually lost his funding. One famous physicist, Richard Garwin, went after Weber and publicly humiliated him. The room-temperature bar detectors Weber favored were discredited as the wrong way to detect gravity waves. The right way was thought to be through a technique known as laser interferometry (LIGO), which of course eventually is how the discovery was made. Weber with his over-blown claims was seen as a danger to the promise of the new detectors. Things turned ugly when Weber in turn attacked the pretensions of LIGO. Weber died in ignominy—a forgotten trailblazer like so many on the losing end of scientific controversies.

That was until 2016. With the new discovery assured, Weber was rehabilitated as a hero—a pioneer in the field. His original apparatus is now proudly displayed alongside the new detectors. As Perillán notes, not all the physicists turned against Weber, Cal Tech theorist Kip Thorne, kept his respect for Weber's achievements during the darkest days of Weber's demise. Thorne was not alone. When I interviewed solar-neutrino pioneer Raymond Davis in the late 1970s, at the height of the campaign against Weber, Davis steadfastly told me that he regarded Weber as a hero—someone who, like himself, knew what it was like to detect something of extraordinary difficulty that flew in the face of theoretical physicists cherished beliefs. But publicly towards the end of his career, there is no doubt Weber was at best a marginal figure and worst a pariah. Once the credibility of a scientist is shot it is almost impossible to recover.

There is no doubt that the eventual detection of gravity waves is a monumental achievement rightly celebrated and one can argue that in the realpolitik world of funding there is no room for a maverick who bucks the consensus—but was the demonization necessary? Even if it was, scientists could be honest about it, and say that this is the way it has to be—an all too human struggle over limited resources. They could even be a little shame-faced about what happened. Why buttress the story as one big happy family on the rocky road to truth, with the hero Weber providing some of the needed foundational rocks?

Perillán shows that even the greatest theories in physics are not immune to the dangers of myth histories—indeed they depend upon them. A notorious muddle conceptually, but indisputable in its ability to enable calculation and prediction, quantum theory was born out of a series of disputes and compromises as to its completeness, how it corresponded to reality, and the very language used to describe that reality. Only recently through the work of scholars such as James Cushing and Mara Beller has the full story been told about how the giants of physics maneuvered to keep together the almost mystical reigning view known after its progenitor, the Danish scientist, Niels Bohr, as the "Copenhagen Interpretation". The well-known particle wave duality was enshrined in the principle of "complementarity", whereby there was no contradiction as long as wave descriptions and particle descriptions of experimental set-ups were kept separate. A consistent account of quantum mechanics was offered by theories such as Louis de Broglie's pilot wave theory which, although requiring complex calculations, kept the theory deterministic. The history that dominant scientists associated with Bohr and Heisenberg told, effaced the controversies over alternative points of view, and de Broglie's theory, like that of many other dissenters, vanished from the mainstream. The aristocratic de Broglie had won a Nobel prize for discovering the wave-like features of electrons but his pilot-wave theory, which he himself ultimately abandoned, is all but forgotten. When new generations of physicists such as David Bohm and John Bell started to probe what had gone before they met unexpected opposition. Mathematical impossibility proofs were trotted out against Bohm and his theories by the quantum elite. Myth-histories is the broad brush name Perillán uses to discuss what ensued. It is arguable that in practice these myth-histories encapsulate a more fine-grained series of moves and issues which science studies scholars have discussed in detail over the last decades. The strength of this book is that it returns us to the big issues at the core of the sociology of scientific knowledge, how credibility is made and unmade in science and at what cost to the overall enterprise. With the "science wars" of the 1990s behind us, this book is a laudable attempt to be consistent with the goals of science studies and also defend science from its own excesses.

It is in the last most extraordinary case study in the book that we see the full dangers of myth histories for the engagement of democratic societies with science. Known as the "L'Aquilla Seven", seven Italian seismologists stand trial for failing to warn their local community about a pending devastating Earthquake. On the face of it such a criminal case seems ludicrous. We all know earthquake science is uncertain. But by excavating this trial Perillán shows there is more to be said. It seems that these scientists did indeed partly bring the law down upon their own heads. They were so in love with the myth of scientific method that they thought local traditions of sleeping outside during the preceding tremors (known as swarms) were not needed. They stressed that seismologists could never make predictions, that the swarms could be releasing pent up seismic energy (a claim that was much disputed)—also their energy at the time was spent attacking a local expert who indeed predicted a massive event but based his prediction upon a buildup of local radon, which they took to be a pseudoscientific claim. In bashing pseudoscience the seismologists arguably took their eyes off the ball. The wider scientific community of course rallied around the scientists in court (shades of Galileo) and the case was eventually dismissed. As with science studies scholar Brian Wynne's famous analysis of how scientists lost the trust of local Cumbrian sheep farmers during the Chernobyl fallout, this case too is a missed opportunity for the scientists to learn a lesson in what science studies scholar, Sheila Jasanoff calls, civic epistemology.

A science without myths may not be possible but surely having identified the problem we can create new myths and stories about a democratic science? The entanglement of these myths with real courtroom drama leaves me with one nightmare. Imagine the unimaginable. Donald Trump and his like are in power for a very long time. The courts turn against science. Our favorite scientists are on trial—the CDC, Dr. Fauci, yes even that friendly epidemiologist from Harvard who is always on the PBS News Hour. Yes, they have made mistakes: no masks then masks, CDC tests that didn't work, varying epidemiological predictions. Yes they were right to bash the Trump pseudoscience of bleach. We rush to defend the scientists—they have done their best against a deadly enemy, they may have been too optimistic, too pessimistic, science is uncertain, knowledge is full of holes. And we all know if left only to the politicians we would be far worse off. But could it just be that the scientists themselves have some blame to shoulder? This nightmare now seems unlikely with a new democratic President and with the values of science once more being reaffirmed. But the danger is still present - science may over reach in response. The myths the scientists have created and live by may come back to haunt them after all.

Acknowledgments

A project like this has deep, sprawling roots. It slowly germinated over years before coming together in its current form. Countless people, seen and unseen, have touched and influenced this book. It is impossible to list every debt of gratitude I incurred along the way, but I have done my best to acknowledge people who have left their mark on my process. If I miss anyone, please accept my apologies, and know that your unheralded efforts are greatly appreciated. Many eyes have read through iterations of this manuscript, but I alone take responsibility for any mistakes that remain.

Although this is not the typical first monograph adapted from a doctoral dissertation, my interest in scientific myth-histories grew directly out of my interdisciplinary study of the de Broglie-Bohm pilot wave interpretation of quantum theory. This book would not have been written without the formative guidance and support of my graduate mentors in both history and physics at the University of Rochester: Theodore Brown, Dorinda Outram, Joseph Eberly, and Nicholas Bigelow. In addition to their early efforts, they have continued to track and support my career, and this project. Professors Brown, Outram and Eberly all read multiple drafts and gave me incisive feedback right up to final submission.

Bumping into Matthew Stanley while crossing a busy West Village street, in 2008, was a moment with real mythological power in my life. Although heroes don't exist, Matt comes closer than anyone I've ever met to embodying that trope. He has been in my corner since summitting Jabal Musa in 1997, long before this book began to take shape--giving selflessly of his time and expertise and being an exceptional mentor and friend. That serendipitous moment on the streets of NYC led to my first academic appointment at NYU's Gallatin School of Individualized

Study, a place that deeply values interdisciplinary discourse. Gallatin was the first place I felt intellectually whole. There, I began to formulate my ideas and discuss my work with colleagues. Conversations with historians of science such as Matt, Myles Jackson, David Kaiser, and Silvan Schweber were particularly formative. The NYC History of Science Workshop hosted by Gallatin is where I presented my first rough sketch, using myth-history as a theoretical framework. Living in NYC also allowed me to connect with the brilliant physicist Pierre Hohenberg, who was profoundly generous with his time and support and whose quip launched my quest for common ground.

At Vassar College I have found my intellectual family and home. Colleagues in the Physics and Astronomy Department and the Science, Technology, and Society Program have adapted seamlessly to having me in constant superposition. They have unflinchingly supported my interests and found the right moments to inject catalytic reflections that moved and shaped this project. In particular, I thank Janet Gray and James Challey for their support, mentorship, and friendship. They have read several drafts of chapters, and generously hosted me in their home on numerous occasions, enduring endless conversations about the project's ebbs and flows. Bob McAulay, too, has been stalwart in his support. His last pass at the manuscript was especially incisive, finding instances of unintended bias and keeping the analysis focused. Christopher White gave his time and expertise generously throughout the writing. His unfailing confidence that this book project would see the light of day was surprisingly contagious.

Chris joined Matt and Nancy Pokrywka for an all-day manuscript review workshop funded by Vassar that was critical to the book's development. A second informal workshop with Bob, Nancy, David Esteban, Jamie Kelly, Christopher Raymond, Abigail Coplin, Wayne Soon, and M Mark, helped prepare the manuscript to present to publishers. As an institution, Vassar has been heavily invested in my success and extremely supportive of my work. Katherine Hite has been an exceptional Faculty Director of Research Development, consistently and creatively marshaling institutional resources to fund academic publishing seminars, manuscript review workshops, and professional editing services for faculty. Thanks to Katie's initiatives, and a helpful push by Carlos Alamo-Pastrana, I've become deeply indebted to Judith Dollenmayer's editorial wizardry; she has helped iron out countless tics and errata and helped me become a more effective writer. I add deep thanks to Joseph Nevins, Debra Elmegreen, and Brian Daly for taking time to read portions of the manuscript and provide invaluable feedback and support during the final stretch.

Since I chose to teach at Vassar because of its students, complete acknowledgments must not fail to credit their role in this project. From the outset, Vassar students have been true collaborators on this book. They have allowed me the space and time to refine ideas through iterative pedagogical revision. Each case study in this book has grown out of coursework in STS and physics classes. All my analyses have been deeply informed by class discussions and student perspectives. It's difficult to express how fortunate I am to teach and learn from Vassar students each day. Working with STS majors on their theses has taught me to wade into new intellectual territory with courage, curiosity, and humility. In particular, I highlight the wonderful work of my research assistants: Mikayla O'Bryan, Annie Xu, Hannah Martin, Joshua Yannix, and Zeyu Liu (Margaret). Working tirelessly under difficult and remote circumstances, Margaret was instrumental in crafting the book's illustrations.

Outside Vassar, I have benefited from a rich network of scholars from various fields. The book has evolved as a result of insightful feedback from countless colleagues at many conferences over the years. It testifies to the power of interdisciplinary discourse in a search for common ground. In one form or another, the case studies in these chapters were presented at the American Institute of Physics (AIP), the History of Science Society (HSS), Society for Social Studies of Science (4S), the British Society for the History of Science (BSHS), and the European Society for the History of Science (ESHS). Chapter 2 appeared as a journal article in *Historical Studies in the Natural Sciences* (HSNS), and I am profoundly grateful to the referees and the editors at that journal, especially David Kaiser and Olival Freire Junior, for thoughtfully shepherding me through the process. Trevor Pinch read an early version of the HSNS paper and has been a generous mentor and advocate since. I'm honored to have him contribute his reflections in the Foreword to this book. In addition, Spencer Weart gave invaluable feedback during the final stretch, and Dennis Delgado generously contributed the richly haunting cover art. The editorial team at Oxford University Press has been outstanding. This book was an unconventional project, but Senior Editor for the Physical Sciences, Sonke Adlung, was able to hear its resonance and became a champion of the project.

Closer to home, I can't say enough about the support of my family and friends. Although listing all of them here is impractical, they know this project would have been impossible without them. They have stood on the frontlines of my vulnerability and obsession over the years, bearing the brunt of consequences a project like this inevitably entails. I owe them everything. My siblings Julio, Lucia, and Pablo Perillán all read portions of the manuscript at various stages, and I would be remiss not to mention that my mother, Dolores Gandarias Perillán, managed to read the manuscript and send me a long list of errata. Madre, some things never change! Although my father was not here to proofread the manuscript, my memory of him is imprinted all over this book. A telecommunications engineer, Luis B. Perillán taught me early and invaluable lessons about signals and noise, sowing this project's first seeds.

Speaking of seeds, one of the most important courses I took as an undergraduate was "Theories of Religion" taught by William Scott Green. Bill's course introduced me to foundational concepts such as the notion of common ground, the power of mythology, Mircea Eliade's unpacking of the sacred and profane,

William James' varieties of religious experience, and Émile Durkheim's exploration of collective effervescence. Beyond these latent seeds planted decades ago, Bill generously gave of his expertise and piercing intellect by closely reading the entire final manuscript and offering critically insightful feedback in the home stretch. The two TAs from that formative course also left their marks on this book project. Matt Stanley (mentioned earlier) and Jorge Rodriguez who read through early drafts of Chapter 2 and has been an invaluable sounding board and friend throughout this whole process.

As it happens, the most important gift from that "Theories of Religion" course was meeting my *media naranja* Rebecca Thomas. Over the years Rebecca has taught me many, many things about myself and about life, but in order to finish this book, I was inspired daily by her courageous perseverance. Words are entirely insufficient to express my love, gratitude, and debt. Rebecca and our two wonderful children, Quino and Eleni, have suffered through too many working weekends, too much exhaustion, and occasional outbursts of frustration. This book is dedicated to them. I can't thank them enough for their love, patience, and unconditional support. Quino and Eleni, get ready, I now have the time and energy to get you that dog I promised!

Contents

Introduction-Reconstructing Scientific Pasts

Tensions and Explorations

The universe is made of stories, not of atoms.[1]
—MURIEL RUKEYSER

Friendly Banter

Graduate work in both physics and history taught me to use highly specialized research methods to rigorously search out truth and eradicate myths. In spring 2012, I brought this mindset with me as I sat down for lunch with physicist Pierre Hohenberg at the Apple Restaurant near Manhattan's Washington Square Park. Pierre was a brilliant physicist and a family friend. Toward the end of his life, he was particularly invested in work on the foundations of quantum theory.[2] My dissertation on the history of quantum interpretation debates naturally piqued Pierre's interest, so we had discussed these topics at length.

Our conversation that afternoon shifted to comparing the development of quantum mechanics with the history of classical mechanics. After lunch, we sat at our table and sparred good-naturedly about history. It was a lively debate, but I sensed Pierre becoming more and more frustrated with my corrections of his historical misconceptions. As the volume and intensity of our talk peaked, he blurted, "José, I don't care if that's not how it actually happened, it *should* have happened that way!"[3]

As a newly minted myth-slayer, I was stunned. Pierre was precise in arguing about the nature of quantum theory, careful to account for every mathematical nuance and scientific principle, yet, at a moment's notice, he seemed willing to brush aside historical facts because they didn't conform to his understanding of how the past *should* have happened. From the perspective of a historian of science, I found this a fundamental asymmetry. How can scientists like Pierre pay utmost reverence to natural facts and their particular contexts, yet be comfortable rearranging and manipulating historical facts to suit their purposes? Unfortunately, I missed the opportunity to thank Pierre directly for his catalytic inspiration, but this book is an attempt to explore and understand his interjection.

The more I reflected on Pierre's perspective, the more I thought about my own uneven introduction to the history of science, first in my studies in physics, then during training as a historian of science. The lengthy anecdotes I heard in physics classes or read in textbooks and popularizations of science left an indelible mark. Reading John Gribbin's *In Search of Schrödinger's Cat* in my first-year modern physics course was nothing short of transformative.[4] For an impressionable eighteen-year-old, the stories of young quantum revolutionaries overturning established knowledge and revealing a new vision of the world brought physics to life. I was captivated and inspired by the stories of scientific heroes boldly debating the foundations of reality, piercing the veil of the unknown. Eventually, experimental work in quantum optics allowed me to frame those revolutionary tales more sharply and gave me a new, firsthand perspective on scientific practice. Indeed, these experiences displayed dimensions of science more akin to Thomas Kuhn's normal science than to revolution.[5] I was impressed by the rigor, ingenuity, and perseverance of my colleagues in the face of the daily grind of physics research, but at the same time disheartened by the inescapably contingent nature of scientific progress.

As I pivoted to graduate work in history, I began questioning many inherent assumptions about science and its past. I still remember the unsettling impact of reading Steven Shapin's opening line from *The Scientific Revolution*: "There was no such thing as the Scientific Revolution, and this is a book about it."[6] Just like that, the clean and tidy progress narratives I had encountered for so long began to cleave and crumble. Carefully researched micro-histories—historical reconstructions contextualized within a particular time and place or fixated on the evolution of a specific concept such as objectivity—challenged my understandings of science and its practitioners.[7] Clear windows into human and social dynamics, missing from traditional progress narratives, were now opened. I was indoctrinated into the guild of professional historians, taught to revere archival documents and unpack them within their proper historical context, doing everything possible to avoid any hint of present-centered "Whiggishness" in my work. "Whig history" refers to a reconstruction of the past that is framed from the perspective of the present to highlight a seemingly inevitable progress narrative. Chapter 1 discusses the stigma associated with this category and its pejorative use by historians of science.[8]

The prohibition against Whig history was drilled deep into my psyche. Framing and interpreting the past from the perspective of the present, as Pierre had done, is a cardinal sin for mainstream professional historians. Thinking about historical episodes as causally connected stepping stones to an inevitable present is a sure way to introduce anachronistic bias into any historical analysis. If historical actors cannot foresee the future, why should we read their work as confidently anticipating certain future developments? Although these are important issues for all professional historians, they are especially problematic for historians of science trying to articulate historical contingency in the face of Whiggish portrayals of science.[9]

A clear demarcation seems to arise from historians' prohibition against Whig history. As a result, historians and scientists approach and understand science's past differently. From the perspective of professional historians of science, the stories scientists tell are judged to be poorly executed histories. They are Whig histories that perpetuate distortive myths about science, to be corrected or eradicated. As such, a treasure trove of carefully crafted historical reconstructions taken from almost a century of scholarship can debunk long-standing myths, such as notions that: astrology and alchemy have always been pseudoscientific superstitions with little impact on the history of science; Isaac Newton's encounter with a falling apple led him to replace God with his law of gravitation; Charles Darwin's theory of evolution destroyed his faith in Christianity; or, more generally, that the scientific method accurately reflects what scientists do.[10]

Yet to many scientists, the Whiggish stories they tell, regardless of historical veracity, are an indispensable part of science. They highlight the exceptional nature of science and its progress, they help teach students about scientific principles, and, in many ways, they form the connective tissue of scientific communities. Simply put, these stories play a social role for scientists akin to traditional mythologies. They create a powerful framework for collectivity and shared heritage that scholarly histories do not.

Establishing Common Ground

Fortunately, I have spent the last decade immersed in a rich interdisciplinary discourse among historians of science, scientists, and scholars of science and technology studies (STS).[11] Years of writing, teaching, and engaging practicing scientists and those who study the social underpinnings of science and technology have pushed me to look beyond myth-slaying and the binary pitting of disciplinary histories of science against professional histories of those same disciplines. Rather, my scholarly interest focuses on the rhetorical analysis of the stories scientists tell. Without falling into a myth-slaying dynamic, I want to better understand the rationale behind the scientists' accounts and their impacts. Out of respect for Pierre, I want to problematize his outburst without dismissing his interpretations of the past.

My goal is to establish a common ground from which to unpack and translate the tensions inherent in divergent perspectives on the history of science. The relationships among these different perspectives are complex and deeply intertwined. Historiography, the philosophy of history, and the study of scientific rhetoric have a long, rich discourse on this point.[12] To find common ground, we must resist the reductionist urge to eliminate competing alternatives or the equally problematic outcome of collapsing multiple interpretations into an epistemologically compromised, incoherent whole. This book's notion of "common ground" unapologetically embraces the constructive tension of epistemological pluralism. As such, it owes much to the debates within feminist philosophies of science around standpoint theory.[13] It's a place that intentionally accommodates multidisciplinary and interdisciplinary discourse and allows various perspectives on the past to stand, while engaging in critical discourse among them. The goal here is for constructive translation and understanding, not normalizing compromise.

I do not intend to build the foundations of this analysis on a common ground of quicksand. Epistemological pluralism is not synonymous with anything-goes relativism.[14] Not all histories are created equal; but, as master storyteller Chimamanda Ngozi Adichie notes, "Stories matter. Many stories matter."[15] Each narrative should be evaluated, but with appropriate metrics. Importantly, metrics like historical veracity are not always most appropriate. When scientists tell stories about the past, they do something wholly different from professional historians. Rather than prejudge them as inferior, this analysis will unpack the tension inherent in scientists using stories. Many scientist-storytellers may well engage in "bad" or "distortive" scholarly history as evaluated by the criteria of professional historians. Yet these criteria are not universal. The stories scientists tell should also be evaluated based on their own metrics. The key to a fertile common ground is to reserve judgment while examining the underlying assumptions and ideological frameworks that undergird divergent historiographic traditions. We all tell stories about the past for many reasons: this book urges us to engage more reflectively with the underlying assumptions and hidden power dynamics that infuse the stories we tell.

Opening an address at Columbia University, Albert Einstein once observed that "science as something existing and complete is the most objective thing known to man. But science in the making, science as an end to be pursued, is as subjective and psychologically conditioned as any other branch of human endeavor."[16] This juxtaposition of seemingly contradictory characterizations of science may be eerily familiar to readers acquainted with STS scholar Bruno Latour's influential textbook *Science in Action*. In it, Latour famously juxtaposed an older-faced Janus as "ready-made science," looking toward the past, with a younger-faced Janus as "science in the making," gazing forward to the future.[17] Almost a century after Einstein's original juxtaposition and more than three decades after Latour's two-faced Janus, the fundamental tension between these two characterizations of science still haunts us, and the stakes could not be higher.

Latour recently sat for an interview with *Science* Insider in which he admitted to having been a young provocateur who channeled his "juvenile enthusiasm" to "put scientists down a little."[18] Yet, in a post-truth world swirling with "alternative facts," he has come to realize that the true "science wars" were not the relatively short-lived disciplinary skirmishes of the 1990s, but the war on science being waged today.[19] I'm not sure there is a formal "war on science," but there is little doubt that the complex "anti-science movement" about which Gerald Holton wrote so provocatively three decades ago continues to erode public confidence in science.[20] This is especially true in the U.S., where, amid the 2020 COVID-19 pandemic, Dr. Anthony Fauci, Director of the National Institute of Allergy and Infectious Diseases (NIAID), decried a growing "anti-science bias."[21] Although no single cause exists for deepening public mistrust, part can be attributed to a gross misunderstanding of what science actually is and how it is practiced.[22]

Things have changed dramatically for Latour, who now finds himself on the other side of the fence, defending scientific authority and advocating for a common reality with commonly understood facts.[23] Even as he now allies with scientists, Latour is careful to note that his positional shift does not change what he sees as the antidote to eroding respect for and authority of science. He believes that scientists need to be more transparent in presenting their science as "science in action." That entails being forthcoming about "their interests, their values, and what sort of proof will make them change their mind."[24]

Latour is not alone in his diagnosis of the problem. Historian of science Naomi Oreskes's *Why Trust Science?* examines the causes of, and possible solutions to, the erosion of scientific credibility.[25] Oreskes convincingly argues that scientists must become better communicators, explaining "not just what they know, but how they know it."[26] How exactly does a scientific community certify something as established scientific knowledge? What are the social mechanisms that allow scientists to judge and translate each other's work? If scholars in other fields, students, and the broader public had a better sense of how scientists do research and engage each other in various forms of social "transformative interrogation," they would be more apt to trust scientists.[27]

Although I agree with Latour's and Oreskes's calls for greater transparency, I wonder if the same criteria should also extend to representations of past science. After all, how we interpret and reconstruct our past profoundly affects how we see ourselves in the present and project ourselves into the future. This book asks this fundamental question by trying to establish common ground between scientist-storytellers and historians of science.[28] Unsurprisingly, the answer is not straightforward. The stories that scientists tell tend to have great rhetorical power but lack historical rigor. Sometimes this lack of rigor comes from ignorance; other times it is a willful, pragmatic leveraging of power. Moreover, although there are examples of scholars who have pursued graduate training in both science and history, it seems unlikely that most scientists interested in the history of their subject will decide to follow this arduous path.[29]

So, should scientists stop telling stories about the past and leave historical reconstruction to professional historians? Not at all. The goal of seeking common ground is to seed a more reflective relationship between scientists and the history of their craft. In carving out room for constructive interdisciplinary conversations, one hopes that scientists become more curious about professional historical reconstructions and leverage them to improve rigor. However, that alone is not enough. Even if scientist-storytellers persist in creating Whiggish tales of past science, they should do so with clearer understanding of their rhetorical power. For their part, historians of science can better appreciate how narratives become social actors in scientific communities. They can also benefit from being more reflective about the inherent tensions between highly contextualized historical details and persistent social phenomena.

Unpacking Einstein's Black Box

If there is any hope of establishing a constructive common ground, we need to unpack the black box at the center of Einstein's two divergent characterizations of science (see Figure 0.1).[30] How does this black box account for the seemingly abrupt transformation between the image of a subjective science in the making (on the left) and a science that is existing, complete, and the most objective thing known to humans (on the right)? If we open the black box, will we find anything inside? Could this transformation be a natural transition triggered by science's exceptional capacity to distill and refine knowledge? Although I don't discount these possibilities, I claim that a careful examination of the stories scientists tell can help reveal the dynamics hidden in Einstein's black box.

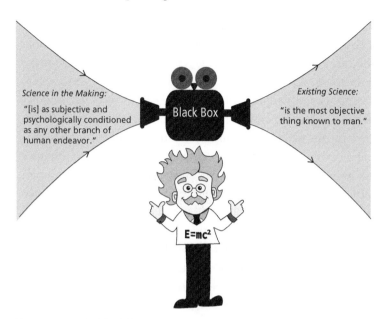

Figure 0.1 The black box in the center seems to transform Einstein's subjective science in the making into an existing science that somehow seems objective and complete. Establishing a common ground requires that we work to unpack this black box. (Source: Illustration by Zeyu [Margaret] Liu.)

Einstein's characterization of science as something "existing and complete" idealizes objective science. It invokes a universal progress narrative that establishes a scientific past infused with normative methods and practices that have worked to unveil immutable truths about the world and a cumulative aggregation of certified scientific knowledge. The past is established as anticipatory, not at all contingent, while the present is seen as inevitable. To paint this portrait, human frailty, methodological errors, and blind alleys are usually filtered out of the characterization. Imperfect human practices within messy, convoluted social contexts are sanitized of doubt, recast into heroic tales in which each successive generation of scientists is causally linked as if it stood on the shoulders of giants. As a projection of communally shared ideals, this characterization of science clearly invokes what social scientists have come to call a "social imaginary."[31] Such imaginaries are not just abstract figments of our collective imagination, they are foundational, malleable frameworks that help us make sense of our world.

The second of Einstein's characterizations directly acknowledges the human and social conditioning that is part of all science in the making. It is grounded in particular contexts that imbue a sense of contingency and uncertainty. It requires that we ask probing questions about patronage, power, the institutionalization of science, and the certification of its knowledge. It problematizes the sense of universalism and continuity that emerges from Whiggish progress narratives. Einstein's two characterizations seem to arbitrarily differentiate the subjective nature of science in the making from its objective, established past. Obvious questions arise from this juxtaposition. If all past science was once subjective science in the making, when and how did it become objective? Succinctly, what's in Einstein's black box?

The fundamental tension between Einstein's characterizations of science is reflected in the disparate uses of science's history—shrouded in diverse methodologies, varying intents, and dissonant assumptions about the nature of both scientific practice and its history. This tension is easy to identify but difficult to resolve. Although part of the controversial back-and-forth about Whig history involves defining or preserving disciplinary identity and autonomy, an overly dichotomized framing misses much subtlety inherent in this tension. As we will see in Chapter 1, due to its inflammatory usage and controversial past, Whig history has become, at times, a polarizing category with diminishing analytical returns.[32]

Myth-History

In order to establish a fruitful common ground, we must preserve and unpack the historiographic tension arising from Einstein's black box. To begin, Chapter 1 explains in greater detail the notion of "myth-history," a category formulated by scientists themselves. The term has been used by physicists to self-consciously distinguish the informal stories they tell from historical narratives written by professional historians. In his popular history of particle physics, the late Nobel laureate Leon Lederman waits until a brief postscript to admit he has just written a myth-history.[33] Lederman is not admitting wrongdoing. Although he is aware of the deficiencies in historical rigor underlying his myth-history, he reserves the right to intentionally filter out historical details for what he considers the greater good of science—and his quest for something other than historical truth.

As Lederman notes, "There may, in fact, be no source for some of the best stories in science, but they have become such a part of the collective consciousness of scientists that they are 'true,' whether or not they ever happened."[34] A professional historian may recoil at this statement, but as a teacher and scholar who straddles the history of science, physics, and STS, I am interested in a more reflective understanding of the rhetorical power and impact of these myth-histories. If these anecdotal stories never happened, in what sense are they "true"? How should we understand the notion of a collective consciousness of scientists? How is it negotiated, and what power structures does it reveal? In what contexts and under what conditions are myth-histories deployed? To what extent are scientist-storytellers aware of their rhetorical power and impact? With questions like these in mind, this book studies the different ways scientist-storytellers use myth-histories, and the consequences of their use.

By enlisting scientists' own historiographic category and establishing a common ground, this analysis seeks to more subtly understand the apparent asymmetry between scientist-storytellers' use of historical and natural facts. The stories scientists tell are not just poorly researched scholarly histories or distortive Whig histories to dismiss; they are myth-histories. The hyphen is critical, for it signals the presence of a chimeric genre that bridges distinct narrative modes. Myth-history does not replace scholarly history; it's an alternate form, a different species.[35]

In communicating their science, scientists tend to use these chimeric narratives for important rhetorical purposes. Myth-histories, like those in textbooks and popularizations of science, employ history as a rough scaffolding. They filter out unwanted historical details and infuse narratives with mythological tropes drawn from prevailing social imaginaries of an ideal science resting on the shoulders of scientific heroes. The stories scientists tell undoubtedly deliver value, coherence, and inspiration to their communities. They are tools used to broker scientific consensus, resolve controversies, and navigate power dynamics. Yet, in foregrounding these imaginaries, myth-histories also spread unintended consequences that must be brought to light. For example, the same cohesive and convergent forces that these narratives bring to a scientific community can also erect barriers to inclusivity, which leave women, people of color, and those with divergent ideas feeling that they don't belong.[36]

Potentially even more damaging to science is the possible effect of these myth-histories on the public's perception of science. The intent of these stories is to inspire confidence in objective scientific progress. However, in an era of twenty-four-hour news cycles, pre-emptive press releases, tweet storms, social media echo-chambers, and limited attention spans, repeated dissonance between projected aspirational ideals of scientific certainty and the real-time uncertainty that accompanies the messy, contingent nature of scientific practice has contributed to eroding public trust. When the political and social impacts of this erosion are being so deeply felt, it is critical to engage in provocative, fruitful discourse on myth-histories' rhetorical power.

In a presidential address to the American Historical Association, William H. McNeill cautioned his colleagues to resist their proclivity for myth-slaying. Reminding them that "eternal and universal Truth about human behavior is an unattainable goal," he thought it better to work toward a "more rigorous and reflective epistemology" that might lead to "a better historiographical balance between Truth, truths, and myth."[37] McNeill got to the heart of the matter. Instead of focusing on mutually exclusive boundary conflicts, we should embrace McNeill's decades-old call for open discourse, greater reflectivity, and a constructive common ground.

All historical writing, no matter how scholarly, is to some extent mediated by interpretative frameworks that select, translate, and curate primary sources for particular audiences. Just as scientific data do not speak for themselves, historical facts also require human interpretation. To analyze history is to interpret and curate it. In particular, the writing of any linear history of science that grapples with the flow of time and attempts to distill a faithful representation of cause and effect will necessarily require some interpretation and subjectivity. More important than disciplinary disputes about who owns the history of science is to examine how myth-histories are constructed, deployed, and metabolized within scientific communities and, more broadly, in society.

Admittedly, in looking to engender an open interdisciplinary discourse, this book emerges from an inherent structural asymmetry. The objects of study will be the stories scientists tell, not histories written by professional historians. This study has no claim to completeness. It is intentionally limited by pragmatic concerns about accessibility and coherence. The hope is that by focusing on the use and impact of these disciplinary narratives within an epistemologically pluralist framework, this book will loosen the vise-like grip of strict segregation and allow constructive discourse. To that end, after Chapter 1, an examination of myth-history as a useful historiographic category, focus shifts to four case studies that illustrate the uses and impacts of myth-history in different contexts.

Roadmap of Case Studies

Outside of the explicit intent and rationale behind myth-histories described by scientists themselves, these narratives tend to have great rhetorical power and social agency that bear unintended consequences. To understand the ways myth-history's social potency manifests, this project examines four case studies to establish it as a useful historiographic category. The case studies are chosen for their range in subject matter and context. Each case study also highlights different aspects of myth-histories' intended and unintended consequences.

As with all case studies, selection bias is a real concern.[38] None of these analyses can pretend to match the depth of a well-researched monograph, but the rich analysis that emerges from weaving together a patchwork of vignettes can be equally power-ful.[39] Doubtless, some details are missing from each examination, yet, wherever possible, I have tried to present a fair analysis. Inasmuch as this book is an invitation to an open, reflective, inter-disciplinary discourse, some readers may strongly disagree with my analyses and conclusions. I welcome them to engage and help build a common ground around myth-history.

Each case study is uniquely situated to answer particular questions about the use and impacts of myth-histories. Together they show the eclectic use of this narrative form in various historical, social, geographical, and scientific contexts. On the one hand, the cases represent various periods and contexts in the history of science over the past century. Although not strictly chronological, they roughly track that way. Along another axis, the studies begin with quantum myth-histories that affect mostly inward-facing community dynamics, and end with a public-facing ana-lysis involving existential tensions arising out of seismic uncertainties.[40] The cases illustrate varying levels of engagement between scientists and the social institutions of which they are a part, and on which they have come to depend. Finally, a third axis shows the variety of scientific disciplines that use myth-histories. Although physicists seem to have coined the category, the case studies examine its use in physics, biology/biotechnology, astron-omy, and seismology to illustrate the wide use and applicability of this form as a rhetorical tool. There is a particular logic and flow to the order of these studies, so it is worth taking a moment to scan the case study roadmap.

An unhappy complaint by celebrated Irish physicist John Stuart Bell, who challenged an unchecked quantum orthodoxy, opens Chapter 2. At first Bell seems little more than a disgruntled physicist blowing off steam about his education. Closer examination reveals much higher stakes. The chapter probes Bell's frustrations about his training in physics at Queen's University Belfast in the late 1940s, which stemmed from entrenched quantum narratives that took hold in the early 1930s and continued to dominate the field for decades. The orthodox quantum interpretation eventually became synonymous with the city of Copenhagen and was used widely in the international physics community to filter out unwanted alternate interpretations, shut down interpretational debate, and promote a pragmatically productive culture of scientific consensus.

This case study comes first not only chronologically, but also because it represents the most scientifically internal discussion of the four. At first, the impacts of these quantum myth-histories seem quite limited. They appear restricted to interpersonal power dynamics within a relatively small community of physicists. Yet, like "ripples in social spacetime," the impacts of these myth-historical narratives transcend their immediate working communities.[41] This case shows how the creation and dissemination of quantum myth-histories contributed to the congealing of consensus with long-term pedagogical and scientific consequences.

Chapter 3's case focuses on a caustic patent dispute over gene-editing technology, waged partly through published myth-histories. The revolutionary gene-editing technology CRISPR-Cas9 has quickly become a vehicle for patent and priority controversies to determine who cashes in on billions of dollars in licensing fees, Nobel Prizes, and scientific immortality. The chapter probes competing myth-histories in the context of larger socioeconomic forces, as well as the absence of an international regime of ethical guidelines for this work.

As this case study shows, the old adage that history is written by the victors requires revision. The adage assumes that histories are written after a winner has been declared. But in Chapter 3, competing histories can also be promotional battlegrounds for would-be victors. The case of CRISPR myth-histories differs significantly from the quantum case study of Chapter 2. Beyond the obvious differences of subject matter and context, there are inherent differences in their topologies of power and breadth of myth-historical impacts. In the context of the CRISPR patent controversy, we can see that myth-histories can have broader social repercussions that pervade legal, bioethical, and economic concerns.

Chapter 4 takes up the critical trope of scientific heroes. In carefully contextualized historical reconstructions, there are no absolute heroes and villains, only humans with competing interests. These are myth-historical categories constructed in retrospect and with all the power and authority of a present-day scientific community shaping its social imaginary.[42] Portraits of scientific heroes like Einstein are drawn from idealized projections of scientific practice that reinforce aspirational community norms and are, effectively, impossible to live by.[43] In myth-historical reconstructions, scientific heroes tend to personify these ideals, and pariahs their polar opposites. However, since all historical figures live in the gray area and fail to follow these idealized norms precisely, scientist-storytellers must often decide ex post facto how to cast their protagonists, as heroes or pariahs.

Not surprisingly, at times the line between these categories is porous and far from static. This chapter looks at one of the clearest recent examples of this porosity by examining myth-historical narratives constructed in the wake of the first direct detections of gravitational waves in 2015–16. These narratives cast Einstein as a caricatured scientific hero, but they also engage in a clear ex post facto rehabilitation of physicist Joseph Weber. This posthumous transformation is quite stunning; Weber's reputation as scientific pariah is quietly replaced by that of a pioneering hero. On the surface, this myth-historical reconstruction seems benign. It can even be argued that the reassessment of Weber is fair and benevo-lent. Yet one can't help wondering if unintended consequences stem from rehabilitating and recasting Weber's role in this cele-brated discovery.

If the discussion in Chapter 2 is most inward-facing, Chapter 5 should be considered the most external-facing of the four cases. It grapples directly with the tensions between idealized portrayals and the public's perception of science. In the early morning hours of April 6, 2009, a devastating earthquake struck the city of L'Aquila, Italy, killing hundreds of people. Many of L'Aquila's residents were not properly prepared for the earthquake and felt betrayed by the scientific and technical experts tasked with public safety. Three years later, in a shocking outcome, a group of seven scientists and public officials were convicted of manslaughter, and sentenced to six years in prison for their actions before the earth-quake. Although the L'Aquila Seven's convictions were eventu-ally reversed or suspended, the controversy illustrates the unintended consequences of scientists and public officials engaging with myth-historical imaginaries while publicly per-forming "boundary work" to demarcate science from pseudo-science.[44] After the 2009 earthquake, many Italians affected by the natural disaster felt betrayed by science. This last chapter unpacks the tension between a scientific ideal and the messy uncertainty that accompanies all scientific practice, as it relates to the effective communication of scientific knowledge.

Emergent Themes from Myth-Historical Case Studies

All four case studies illustrate scientists employing myth-histories and their social imaginaries to help communicate their science to students, colleagues, or the broader public. Scientist-storytellers are participating in an important rhetorical tradition in science. The stories they tell help to reinforce existing ideals of objective scientific practice that do not necessarily correspond to the scholarly histories of science written by professional historians. We find myth-histories used in a variety of circumstances and for various internal and public-facing rhetorical purposes. Yet, in all cases, they mask the fundamental tension hidden in Einstein's black box: between science as a socially embedded activity and science as a rational endeavor, between science in the making and science existing and complete. To comprehend the causes of this tension, we must better understand the rhetorical power of myth-history.

The impacts of myth-histories in these case studies can be grouped into three broad, interrelated categories: science pedagogy, consensus work, and the public understanding of science. Obviously, not all cases show all forms equally clearly. As we navigate through the case studies in Chapters 2–5, we see situations in which the use of myth-history illustrates one of these categories more directly than the others. Yet, as these cases also show, to some extent the ripples in social spacetime emanating from the use of myth-histories eventually permeate all categories. Intended and unintended consequences can be interpreted as positive or negative, based on the reader's perspective. So, whenever possible, these analyses try to honor the epistemological plurality of the established common ground and contextualize any unavoidable judgments.

A 2016 study of New York City high school students tested the effects of different historical narrative styles on the motivation of science students.[45] The authors concluded that reading traditional textbook myth-histories about famous scientists reinforces the deep-seated belief conjoining "success in science" and "exceptional talent." Importantly, this belief can negatively affect some students' motivation to learn science.[46] According to the study, reading a myth-historical narrative about scientific heroes that portrays them as flawless solitary geniuses that inevitably revolutionize science can negatively affect the motivation of students in science classes. For many students who struggle to "pull themselves up by their bootstraps" and solve problems on their own, this can clearly signal that they don't belong in science; but these

solitary scientist portrayals hide the social and dialectical foundations so critical to all scientific inquiry.

This study also underscored that not all students respond to
historical narratives in the same way. For some, reading myth-
historical accounts of scientific heroes and icons inspires them to
succeed. Learning from idealized myth-histories can teach students about the community's norms and aspirations. However,
other students who learn about these scientific heroes practicing
idealized science come away feeling that they themselves don't
belong in science. It isn't surprising to discover that students who
responded best to the myth-histories were those who were
already performing well in the class. Students who had struggled
saw their motivation and performance positively affected by the
use of alternate "struggle narratives." These stories of the same
famous scientists talked of personal and professional challenges
they faced along the way.[47]

How should we interpret this study? Should scientists continue
to tell myth-histories that tap into social imaginaries, emphasize
scientific ideals, and inspire the top students in their classes? Or
should they acknowledge that past scientists cannot be fully
understood unless placed in their proper historical and social
contexts, thereby humanizing and making them more vulner-
able? There is no clear normative answer. That said, the myth-
historical versions of science introduced in textbooks, lectures,
and popular accounts should be understood as powerful rhetorical devices distinct from scholarly histories and with the capacity to affect students in a variety of ways. With that in mind, it
behooves scientists to be more reflective in choosing what historical narratives to employ—and perhaps more explicit about
what they hope to achieve by using them.

Beyond pedagogy, myth-historical impacts can also be seen in
the research agendas of scientific communities and broader public
discourse about science. There is little doubt that part of forming
a scientific consensus involves constructing a corresponding
historical narrative that supports the winning ideas, while filtering

out and discarding ideas that don't match the emerging consensus. Richard Staley's study of participant histories argues that "forgetting is integral to scientific advance," yet should not limit our understanding of scientific practice and its historical development.[48] While I agree with Staley, in order to drive consensus, some scientists actively employ alternate myth-historical narratives as rhetorical agents of omission. When this happens, we should recognize narrative construction not as a passive act of forgetting, but an active act of rhetorical filtering. It behooves scientists to ponder what their stories omit. Acts of rhetorical pruning are not just clearing the path of scientific progress; they may also erase important interrogations that may have been prematurely shut down.

For example, the story of the development of quantum physics in the first half of the twentieth century has been told and retold, especially by physicists who participated in the quantum revolution. Yet, as historian of science Mara Beller has noted, instead of simply accepting these orthodox narratives at face value, we can engage in historical analyses that show how revolutionary stories are constructed, "how division between 'winners' and 'losers' is fabricated, how the opposition is misrepresented and delegitimized, and how the illusion of the existence of a paradigmatic consensus among participants is achieved."[49]

Although rigorous historical analyses, like Beller's, can work to unveil "revolutionary stories" as illusions of "paradigmatic consensus," they can also work to reveal the hidden social impacts of these myth-histories.[50] As we shall see in the case studies, myth-historical narratives are powerful social agents that help carve out and protect scientific imaginaries and dictate research agendas. They do this by helping to rationalize what the community recognizes as acceptable scientific inquiry versus activities that may be labeled illegitimate or pseudoscientific. As such, myth-histories are employed by scientist-storytellers when engaging in boundary work. They are used to protect scientific purity by stripping out unwanted social and human biases.

As these myth-historical narratives diffuse beyond scientific communities, they also deeply affect the public's understanding of science. As mentioned earlier, many scientists and their allies sense that the scientific community is now under siege, in a war on the credibility of science. One cause for greater public mistrust is a fundamental misunderstanding of what science is and how it is practiced. Myth-histories have contributed to this misunderstanding by foregrounding their imaginaries and painting caricatured pictures of past scientific practice as objective and detached. With an unrelenting and unforgiving capacity to exchange information from an endless stream of sources, modern technology has made it inevitable that people will be exposed to a more "subjective and psychologically conditioned" picture of scientific practice.[51] The resulting dissonance between these divergent portraits can foment uncertainty and distrust.[52]

When teaching the history of science via myth-histories, scientists are often so confident in their cumulative scientific progress that they present current scientific knowledge as unassailable. For those not engaged in science, scientific practice becomes the act of removing uncertainty and finding absolute truth about the natural world. Stale adages, such as scientists passively reading "the book of nature" or letting data "speak for themselves," leave people with the impression that science is simply a tool for polishing and refining a picture of an objective universe that is quickly approaching full certainty. In fact, scientists understand that their work, while answering important questions about the universe, tends to create more questions and is always wrapped in uncertainty. Although scientific knowledge is expanding, scientists should be more transparent and nuanced in portraying their craft and their relationship to its inherent uncertainty.

If the ideals of science seen in myth-histories point to science producing certainty, any semblance of uncertainty arising in scientific practice can be interpreted as due to scientific failure. The illusion of certainty typical in myth-histories can create impossible standards that ultimately undercut scientific authority.

Repercussions can be dire. In a turbulent social climate, dissonance between illusions of a scientific ideal and its everyday practice are being leveraged for political purposes.

Many academics regard the science wars of the mid-1990s as a fruitless, toxic, and overblown detour of our past, yet in many ways the underlying tension remains unresolved. It echoes the fundamental tension at the heart of Einstein's black box. Although it has been almost a quarter century since the Sokal affair, scientists continue to push back challenges they consider hostile to the integrity of their scientific imaginaries. In many ways, tensions have worsened as these challenges have migrated from abstract academic debates to outright hostilities from a public that increasingly depends on yet distrusts science when it conflicts with their personal worldview.

I believe that science is the most powerful human invention for analytically reducing and studying the natural world. I also believe that science is progressively becoming more powerful in its ability to change the world. However, the metrics we use to judge those changes are not universally given. Rich, reflective, and transparent interdisciplinary discourses that reveal the social underpinnings of science can contribute to restoring mutual trust. An exploration of scientific myth-histories can help lay the foundations of a common ground from which prevailing scientific imaginaries can be better understood and potentially transformed.

Knowing that the communication of scientific content and its practices is critical to the social standing of science, *Science Between Myth and History* argues that scientists should not abandon writing about their history but instead take more care with the stories they tell and the images of science they project. At bottom, manifestations of the essential tension hidden in Einstein's black box are easy to identify but difficult to resolve. This conundrum will stand until we establish a common ground among science communicators, practitioners, and scholars that allows a deeper understanding of the rhetorical power and impacts of scientific myth-histories. This book is an effort in that direction.

1

Myth-Historical Tensions

Origins of a Narrative Category

One of the oddities of history is that legends oftens upersede facts.[1]

—ERROL MORRIS

The Flying Ashtray Argument

A large, heavy, cut-glass ashtray with sharp edges "came hurtling across the room spewing butts and ash."[2] To a young Errol Morris it seemed like the ashtray was "its own solar system" of unfiltered Camel and True Blue cigarette butts flying toward him in an "interstellar gas" cloud of ash. He saw the arc.[3] "Was it thrown at [his] head?" He couldn't be sure, but Morris does remember it was thrown in his direction... "with malice."[4] His time as a first-year graduate student at Princeton University's Program in History and Philosophy of Science came to an abrupt end at the hands of his distinguished professor, the world-renowned scholar Thomas Kuhn.[5]

Morris was profoundly shocked by the dissonance between Kuhn's violence and its serene academic surroundings. "Wait a second. Einstein's office is just around the corner. *This is the Institute for Advanced Study!*"[6] During Morris's short stint at Princeton, Kuhn taught a "two-part seminar on nineteenth-century theories of electricity and magnetism." Apparently, Morris disliked Kuhn's teaching style and his general approach to the history of science. In particular, Morris remembers, Kuhn seemed fixated on the perils of "Whiggishness," complaining endlessly about the distortive effects of presentism on the history of science. According to Morris, Kuhn had become so dogmatically opposed to any form of Whig history that he had "weaponized" an extreme avoidance of all anachronistic language and concepts. He had elevated this strategy of avoidance "to a fundamental principle in the history of science."[7]

Morris recalls the "flying ashtray" argument as the culmination of a caustic debate sparked by a term paper he wrote on Maxwell's displacement current. Kuhn had called Morris into his office to discuss his student's Whiggish interpretation of the history of nineteenth-century electricity and magnetism. Morris interpreted Kuhn's critical feedback as ad hominem attacks. The encounter quickly escalated; Morris called Kuhn a "megalomaniac" with a God complex. Apparently, this was too much for Kuhn to bear. After repeatedly muttering, "He's trying to kill me," the glass ashtray came flying.[8]

How could a debate about the relative merits and Whiggishness of a term paper result in such an existential confrontation? Are we to believe Morris's account, or has this legend of a flying ashtray somehow managed to "supersede facts"?[9] Unfortunately, there were only two witnesses to this encounter, and one of them is dead, so there is no way to corroborate the details of Morris's account.[10] Yet, regardless of its veracity, almost half a century later, Morris's vivid retelling of his ordeal is a reminder that the fundamental tension inherent in Einstein's black box between divergent characterizations of science is still very much a contested boundary and controversial flashpoint.[11]

Unpacking Historical Tensions: Who Owns Our Scientific Past?

In his latest book, *To Explain the World*, Nobel Prize–winning physicist Steven Weinberg reflectively highlights a boundary conflict between scientist-storytellers like himself and scholarly historians of science:

In telling this story, I will be coming close to the dangerous
ground that is most carefully avoided by contemporary historians, of judging the past by the standards of the present. This is an
irreverent history; I am not unwilling to criticize the methods
and theories of the past from a modern viewpoint.[12]

Weinberg is acutely aware that his presentist and "irreverent" history is considered problematic by professional historians of science. Yet he does not shy away from his approach, but
unapologetically doubles down on it. Weinberg recognizes that
judging "a past scientist's success by modern standards" may not
be helpful to historians trying to understand how things actually
happened, but to someone interested in understanding "how
science progressed from its past to its present," it's actually "indispensable."[13] To scientist-storytellers like Weinberg, there is real
value in what many professional historians of science deride as
"Whig history."[14]

Most explorations of Whig history begin with the celebrated
historian Herbert Butterfield's 1931 declaration that the "Whig
interpretation of history" emphasized "certain principles of progress in the past" while producing "a story which is the ratifica-
tion if not the glorification of the present."[15] In problematizing
Whig history, Butterfield claimed that this form of presentism
was "the source of all sins and sophistries in history," the root
of anachronisms, and the essence of "unhistorical" writing.[16]
According to Butterfield, these unhistorical narratives tend to
selectively translate past events in order to make them comprehensible to a present-day audience and are carefully curated to
ensure a logically consistent chain of events that make the present
a justified, indeed inevitable, outcome of the past.

Clear examples of Whiggish history can be found in most
science textbooks and popularizations of science.[17] Open one,
and you will likely find sporadic sidebars or commentaries
dedicated to telling stories about past scientific developments in
the history of science. These bare-bones anecdotes tend to be
brief excursions intended to complement and reinforce the

teaching of scientific concepts and social imaginaries.[18] Scientist-storytellers curate and link entertaining anecdotes about past scientific heroes engaging in what appear to be idealizations of scientific practice.[19] For example, stories about Galileo Galilei carrying weights to the top of the leaning Tower of Pisa to perform experiments on gravity are seamlessly woven together with anecdotes about a falling apple inspiring Isaac Newton to generalize the concept of gravity from terrestrial domains to the heavens above. Linking stories in this way creates the impression that scientific progress is an inevitable chain of progress built on revolutionary moments of genius-fueled insight delivered by solitary heroes who simply light the way forward, read the book of nature, and unveil its underlying truth. Each step along this teleological path brilliantly arises as a consequence of the last, taking us ever closer to a complete and final understanding of our universe.

For decades, Weinberg has argued for the exceptionalism of scientific progress as "something objective," while explicitly connecting it to his use of Whig history.[20] He claims that historians, fearful of anachronism, fail to learn all they can from the history of science. They prefer to focus exclusively on contextualizing the scientific past. In avoiding Whiggish judgment, they end up missing important historical clues. As a result, Weinberg argues that "the Whig interpretation of history is legitimate in the history of science in a way that it is not in the history of politics or culture, because science is cumulative, and permits definite judgments of success or failure."[21] For Weinberg, this conclusion has deep implications. From his perspective, the evolution of scientific knowledge is exceptional and should therefore be set apart:

> If we think that the discoveries of science are flexible enough to respond to the social context of their discovery, then we may be tempted to press scientists to see nature in a way that is more proletarian or feminine or American or religious or whatever else it is we want. This is a dangerous path, and more is at stake in the controversy over it than just the health of science.[22]

Here lies the heart of the matter. For Weinberg and others, this debate is nothing short of an existential crisis. Learning about the history of science in a way that allows any direct sociocultural influences on our scientific knowledge might challenge the very foundations of a science that strives for ideal notions of objectivity, and puts us on the edge of a very slippery slope of subjectivity. It appears that, to Weinberg, the result of all this historical contextualization might be the undercutting of scientific authority and possibly the destruction of our modern way of life.

Given such stakes, one sees why unraveling Einstein's juxtaposed characterizations of science is so important. Historians may judge Einstein's subjective forms of scientific practice incongruous with the claims of completely objective knowledge produced. On the other hand, scientist-storytellers like Weinberg interpret Einstein's black box delineating these two characterizations as a necessary protective boundary that preserves the exceptionalism of science. Weinberg's "boundary work" is an important rhetorical activity to demarcate science from all other human activities.[23]

In response to Weinberg, celebrated historian and sociologist of science Steven Shapin reviewed *To Explain the World* under the title "Why Scientists Shouldn't Write History." His review dissents from Weinberg's Whiggish historical narrative by claiming that historians of science may "express bemusement at Mr. Weinberg's insistence that science advances by rejecting teleology, even as he depicts its history as a triumphal progress from dark past to bright present."[24] Many professional historians agree with Shapin that a Whiggish interpretation of history such as Weinberg's is self-serving and problematic.

Almost a quarter century ago, Shapin published a landmark book questioning the legitimacy of the "scientific revolution" as an objective historical reality.[25] His book opens with the pithy statement "There was no such thing as the Scientific Revolution, and this is a book about it."[26] Weinberg, while acknowledging that there is reasonable evidence against the common understanding of this historical category, claims, "Nevertheless, I am

convinced that the scientific revolution marked a real discontinuity in intellectual history. I judge this from the perspective
of a contemporary working scientist." As a physicist, Weinberg
has no problem accepting that the use of different lenses will
produce divergent images of the same object of study. He does
not claim normative historical authority, but doubles down on
the idea that a Whiggish historical lens produces a legitimate
interpretation of the history of science. In a clear jab directly at
Shapin, Weinberg concludes: "There *was* a scientific revolution,
and the rest of this book is about it."[27]

In light of Weinberg's clear dig, it's not surprising that Shapin
critically reviewed Weinberg's book. But professional historians
of science should be careful not to overinterpret Weinberg. He
does not claim that his historical account of the rise of modern
science is more accurate than those produced by professional
historians. He sees scholarly history and Whiggish history as two
different lenses through which to interpret the past. From
Weinberg's perspective, scholarly histories give more accurate
accounts of what actually happened during particular contextualized episodes, but if one's goal is to understand cumulative
scientific progress, a Whig lens is a better choice. Weinberg makes
judgments on the past as a "contemporary working scientist"
because he considers it the best tool for understanding the
unique evolution and progress of science. In that sense, he seems
well aware of his rhetorical choices. However, it is not clear that
scientist-storytellers like Weinberg fully understand the underlying assumptions and impacts of using their powerful historiographic lenses.

Meanwhile, I wonder, is it possible that historians like Shapin
are quixotically charging at windmills and calling them giants?
What if Shapin changed the title of his review to "It's OK for
Scientists to Tell Whiggish Stories as Long as They Don't Claim
Them as Scholarly History"? This absurdly long title drives at the
fundamental tension around this conflict. Whig history doesn't
replace scholarly history; it's a different analytical lens, an

alternate narrative form, and a different species of history.[28] As long as everyone acknowledges that these different takes on the past are orthogonal activities, professional historians and scientist-storytellers need not exhaust themselves in a mutually exclusive tug-of-war. Instead, we can all become epistemologically nimble enough to navigate between different historical lenses without succumbing to poisonous controversy.

The term "Whig" should be more than an adjective that means inaccurate or distortive. Unfortunately, as evidenced by the flying ashtray argument, the concept has a long, messy history that makes change extremely difficult. Yet, in studying alternative histories such as Weinberg's, one realizes that they are so much more than informative timelines or comic anecdotes used for rhetorical effect. They are powerful tools for social change, tools that leave their mark on scientific communities, the science they produce, and public perceptions of science. Whig histories help blaze a clear path of progress, paint scientific imaginaries that showcase idealizations of its practice, frame acceptable questions of inquiry, and filter out unwanted or marginalized ideas. These stories can also inspire some students while stunting the motivation and development of others. All in all, the legacy of these stories is uneven and difficult to generalize, but their clear impacts beg further examination.

Is using Whiggish history of science fundamentally problematic? What follows explores this difficult question, without offering a definitive answer. Contentious disputes such as those between Weinberg and Shapin are momentary outbursts that manifest long-unresolved tensions about different uses of scientific history. They echo Einstein's black box, which juxtaposes subjective science in the making with an objective science that is existing and complete. Although this boundary conflict is not often reflectively acknowledged or discussed, there is much at stake. Ultimately, how our scientific past is used and portrayed has significant real-world implications that affect how science is perceived, taught, and practiced.

Finding Interdisciplinary Common Ground

As we have seen, much tension seems to hinge on the question of who owns the history of science. The straightforward answer might be professional historians of science like Shapin. However, much of what is actually consumed as history of science is written by scientist-storytellers like Weinberg, and judged to be Whiggishly distortive by professional historians. As Shapin recently lamented in reviewing Weinberg's book, he and his colleagues do not even "own [our] subject."[29] Establishing a common ground and including individuals with competing claims to our scientific past can show how controversies about various uses of history are more than simple disciplinary disputes between scholarly historians and scientist-storytellers. These tensions might be better described as interdisciplinary conversations that desperately need clear facilitation and translation. By developing a common ground of understanding among these different historical frameworks, we can encourage fruitful discourse on the relative merits and weaknesses of engaging with our scientific past as we do.

A wonderful example of the potential of interdisciplinary discourse can be seen in a recent collaborative essay, "Two Kinds of Case Study and a New Agreement," coauthored by physicist Allan Franklin and sociologist of science Harry Collins.[30] In a disarming conversational tone, the two reflect on their decades-long dispute over Joseph Weber's legacy within the physics community as a result of his early controversial claims of gravitational wave detection. In Chapter 4 of this book we will examine the particulars of Weber's claims and their reception. However, to illustrate the power of common ground, I want to briefly draw attention to Franklin and Collins's collaborative essay. Like the various refracted realities in Akira Kurosawa's masterpiece *Rashomon*, Collins and Franklin's description of their divergent accounts allows us to see the complex, layered nature of historical truths.[31] Like the four protagonists in *Rashomon*, Collins and Franklin take turns reflecting on their subjective perspective on past events. What emerges is a rich dialectic that doesn't resolve their differences but does establish common ground.

As a sociologist of science, Collins has spent the better part of five decades deeply immersed in the gravitational wave detection community, conducting extensive fieldwork and writing assiduously about the ripples in social spacetime he has observed.[32] His analysis has highlighted such inherent biases as "experimenter's regress" and the overall contingency of a socially constructed scientific consensus.[33] As a practicing physicist, Franklin began writing an account of Weber's case in response to Collins's sociological work on the topic. His approach focused largely on examining the "published sources" and assuming that scientific consensus was the inevitable result of "rational processes."[34] Initially, exchanges between the two were quite hostile. As part of the growing toxicity of the mid-1990s science wars, they "found themselves insulting each other." Of late, the "violence has gone out of the debate" and their relationship has evolved into one of mutual respect and cooperation.[35]

This positional shift was not due to any clear rhetorical victories on either side. Rather, it was a result of establishing common ground on key epistemological points and accepting that methodological differences might lead to varying but mutually informative descriptions of the past. As we will see in Chapter 4, Collins and Franklin still disagree about much of what led to Weber's loss of credibility within the gravitational wave detection community, but, as Franklin notes, their perspectives can now be seen as complementary instead of adversarial. Reading divergent accounts of historical episodes, one can get "a better and more complete picture of the practice of science."[36] This is precisely the power of standpoint theory and pluralism described in the Introduction (see Figure 1.1). It is also a powerful way to avoid what Chimamanda Ngozi Adichie has referred to as "the danger of a single story."[37] In their new collaboration, Collins and Franklin have managed to preserve and transcend the fundamental tension at the core of Einstein's black box juxtaposition. This is the type of discourse this book aims to advance.

Standpoint Epistemology

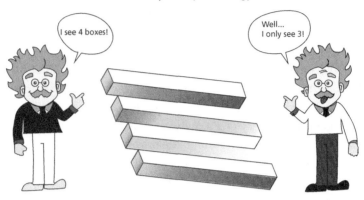

Figure 1.1 Standpoint epistemology. We see Einstein looking at the same object from different perspectives, and seeing something different. From one side he sees four boxes; from the other only three. This reminds us that there is no such thing as unmediated reality; therefore understanding someone else's standpoint is an important precondition for establishing common ground. In their new collaboration, Collins and Franklin have engaged in this powerful form of epistemological pluralism. (Source: Illustration by Zeyu [Margaret] Liu.)

In their essay, Collins and Franklin claim that having transcended the caustic rhetoric of disciplinary boundary work, they are now engaged in what they consider a "normal disagreement between historians rather than mutual incomprehension." Clearly, this transformation testifies to their personal generosity, but it is also a testament to a shifting intellectual landscape, something they term revolutions in both "historiography" and "science studies."[38] Collins in particular seems convinced that the academic world has fundamentally changed and that "bringing the social into the nature of science" is now "treated as a matter of course."[39] It is worth taking a moment to examine this point.

Shifting Historiographic Landscapes

The old caricatured disciplinary conflict framework on display between Weinberg and Shapin threatens to overshadow important reflective work by professional historians and other scholars of science that grapples with the fundamental tension in Einstein's black box.[40] The question of historical objectivity has been a foundational point of stress since before the professionalization of history in the nineteenth century. Yet over the past four decades, historians have been active participants in broader intellectual movements that problematize the notion of objective truth.[41] It is now widely accepted among professional historians that there is no such thing as a single "correct" interpretation of the past.[42] All historical writing, no matter how academic and scholarly, is understood to be mediated by interpretative frameworks that select, translate, and curate primary sources for particular audiences. As Dorinda Outram notes in *Four Fools in the Age of Reason*, the notion that detachment and distance are somehow prerequisites for writing scholarly history "smacks of hubris."[43]

Historians reconstruct the past by representing a particular vision that makes historical "complexity comprehensible" to themselves, and then to others.[44] In the same way scientific data do not speak for themselves, historical facts also require human interpretation. This makes their meaning and relevance subjective. All historical reductions are reconstructions that reflect a particular agenda. The writing of linear history that grapples with the flow of time and attempts to distill a faithful representation of cause and effect will necessarily require some level of presentism. In that sense, these shifting historiographic landscapes have taken scholarly history closer to Michel Foucault's notion of "genealogy" and by extension Friedrich Nietzsche's "effective" history.[45] Yet, for most professional historians, this newfound perspectivism is not the equivalent of anything-goes relativism. Historians continue to work diligently and employ rigorous research methods to ensure that their reconstructions are accurate and faithful accounts of the past.[46]

It is important to note that, although the history of science emerged from academic history, they are not equivalent disciplines. Whether one dates the origins of the modern pursuit of scholarly history directly to Leopold von Ranke or more diffusely within German academia, it is widely accepted that history as a modern scholarly pursuit dates back to the early-mid-nineteenth century. The discipline classified as "history of science" is a much younger subfield. The History of Science Society (HSS) was founded in 1924,[47] but the field did not really find secure institutional footing until after World War II.[48] Even then, it has consistently struggled to develop a coherent, autonomous institutional identity like its "perennial doppelganger, art history," which carved out a clearly delineated departmental home in academia by the 1930s.[49]

Unlike the field of history, history of science has a long tradition of finding itself within interdisciplinary contexts. In some instances, historians of science have found themselves as specialists in history departments; at other times, they have been paired with philosophers of science to form combined programs or become affiliated with STS programs. In a few instances, historians of science have managed to secure their own autonomous departments. Such contextual variations have sometimes given the history of science a feeling of muddled "institutional heterogeneity."[50] The lack of a firm institutional niche led Kuhn, in his 1991 Rothschild Lecture, to suggest that it was a clear sign of disciplinary failure.[51] This "pieced together" and muddled "collage of approaches" led distinguished historian of science Charles Rosenberg to directly question the field's coherence.[52]

More recently, Ken Alder has convincingly argued that by broadening the discourse, this "institutional heterogeneity has kept [the history of science] intellectually supple and perennially reflexive."[53] Yet one may wonder, has this institutional heterogeneity also made historians of science more defensive of their expertise? Does their lack of defined institutional inertia force them to confront scientist-storytellers head on? Do they become a little too aggressive in promoting the importance of context and history? Although these intradisciplinary questions are very interesting and worthy of interrogation, the goal here is to show that the disciplinary conflict between scientist-storytellers and historians of science should be understood within a broader context of historiographical flux.

By the early 1980s, to be labeled a Whiggish historian of science was tantamount to ridicule and ostracism. Yet amid the discipline's aversion to Whig history, some have sought to neutralize this charged dynamic by writing about the benefits of a presentist Whig approach to the history of science and pointing out the inherent problems in clearly delineating and evaluating scholarly or "research" histories against more popular "narrative" histories.[54] As a result, a fascinating historiographic discourse on varieties of historical narrative has grown more fluid than long-standing, polarizing controversies suggest.

Some scholars have tackled the term "Whig history" head on and tried to analyze it more deeply. Others have avoided the term altogether, instead studying characteristics such as anachronism or presentism. Still others have recast the controversy as a broader conversation about the use of alternate communal self-histories like orthodox narratives, pilgrims' tales, participant histories, and official histories.[55] Unlike Whig history, these descriptive categories are mostly divorced from the stigma and historical baggage that might elicit the types of feelings that could result in an ashtray argument. They are more neutral categories that allow scholars to begin to probe the rhetorical effects of alternate historical narratives while avoiding the more controversial boundary conflicts.

Even after a rich intradisciplinary historiographic discourse, we are left with underlying disciplinary tensions between scientists and professional historians that from time to time erupt in public outbursts. Why have conversations, like Collins's and Franklin's about the use of historical narrative, not been more widely adopted? In part, the answer rests on the lack of a common ground, a framework in which neither side feels existentially threatened. This common ground framework, which can allow us to translate across multiple disciplinary intents and methodologies, requires a category that is accessible and intelligible to both scientists and historians. As Collins suggests, we need to find "a shared language and interactional expertise" so we can work together and build mutual understanding.[56]

Since Collins was able to transcend the hostilities and "violence" of his historiographic dispute with Franklin, it should come as no surprise that in one of his earlier books, coauthored with Trevor Pinch, the authors discuss a possible typology that categorizes different forms of historical narrative. In an Afterword examining their interactions with scientists as a result of their controversial book, *The Golem: What You Should Know about Science*, Collins and Pinch point out that there are at least six distinct types of history of science. They categorize these as textbook history, official history, reviewers' history, reflective history, analytic history, and interpretive history. This example of a typology of histories is important for understanding that not all historical reconstructions can be evaluated by the same criteria.[57]

If one could categorize different historical narratives via this typology and then discern evaluative criteria for each, it might result in a powerful analytical framework for understanding the impact of these narratives. This is an excellent example of the multidisciplinary historiographic discourse that has changed intellectual landscapes. Unfortunately, as it stands, Collins and Pinch's typology seems somewhat limiting. Most historical narratives written by scientists tend to be a hybrid of more than one of these types. The variety of intents and contexts that arise in their historical reconstructions makes one wonder how this typology might help us better understand the intent of a scientist-storyteller. Perhaps we need a single, more adaptive category?

Fortunately, a new category need not be invented. Maybe the stories scientists tell should be evaluated by their own criteria. Myth-history is a historiographic category formulated by scientists themselves. Physicists have used the term to self-consciously distinguish the informal stories they tell from historical narratives written by professional historians. To understand the relevance and importance of this category, it is helpful to go back to New York City, circa 1985—a place and time where the idea and power of myth cross-pollinated into historiographic and scientific realms.

Myth-History: Origins of an Interdisciplinary Framework

On a windy February night at Manhattan's Gramercy Park, the "hearty and robust" eighty-year old master of mythology was honored for his contributions to the literature of mythology: "'No one in our century—not Freud, not Jung, not Thomas Mann, not Lévi-Strauss—has so brought a mythic sense of the world back into our daily consciousness."[58] That is how James Hillman introduced Joseph Campbell during a ceremony in which the great mythologist was awarded the National Arts Club's 1985 Medal of Honor for Literature.[59] Later that same year, mythology was so much in people's daily consciousness that two heavyweights in their respective fields, history and physics, would highlight the link between myth and history.

On December 27, the president of the American Historical Association, William H. McNeill, gave his presidential address at the Association's annual meeting in New York City. He challenged his colleagues to bridge the chasm between myth and history. The title of the paper drawn from McNeill's address points to the obfuscated tension between these concepts within the practice of history: "Mythistory, or Truth, Myth, History, and Historians."[60] He begins by pointing directly to the common understanding by historians that "myth and history are close kin inasmuch as both explain how things got to be the way they are by telling some sort of story. But our common parlance reckons myth to be false while history is, or aspires to be, true."[61] Professional historians see their work as a rigorous, methodological, and scientific approach to understanding the past as accurately as possible. On the other hand, they generally interpret myth as involving human subjectivity, imagination, and distortion of the past. As a result, we often see historians justify their work by claiming to be dispelling or slaying one or another myth.

McNeill acknowledges that historians work with relative truths and have no access to absolute historical Truth. In fact, he points out that what might be understood as truth to one historian may be myth to another. Arguments like McNeill's were important contributors to the fluid, reflective historiographic discourse discussed earlier.[62] With the backdrop of this shifting historiographic landscape, his intervention can help us begin to chart a difficult course through the boundary territory of myth and history. Even as McNeill accepts that there is no such thing as absolute Truth, he does not take this premise to its relativist end, holding out hope for his own and his colleagues' intellectual pursuit of a higher historical truth.

For McNeill, it is clear that all truths are not created equal. To address the internal conflict between a fragmented, postmodern epistemology and a discipline that purports to strive for historical narratives that accurately and faithfully represent past events, McNeill proposes to alter the very foundations of modern historiography by developing a

> more rigorous and reflective epistemology, [with which] we might also attain a better historiographical balance between Truth, truths, and myth. Eternal and universal Truth about human behavior is an unattainable goal, however delectable as an ideal.[63]

Unfortunately, details of McNeill's new "rigorous and reflective epistemology" were not fully articulated in his presidential address, but it is clear from his public comments that he believed the relationship between mythology and history could not be reduced to a simple polarizing conflict.

Three years after McNeill's presidential address, his University of Chicago colleague Peter Novick published a landmark study of American professional history's long, evolving engagement with what he terms the "objectivity question." Novick situates McNeill's challenge to historical truth within the broader context of an extended, tortured dynamic in which professional historians have, for long stretches, fetishized "historical objectivity." He is quick to point out that the notion of historical objectivity is not a singular concept, but "a sprawling collection of assumptions, attitudes, aspirations, and antipathies."[64] As his study shows, it's a contested category with a history as old as the professionalization of history itself. The turbulent and contentious boundary between objectivity and relativism points to long-standing intradisciplinary tensions that defy simple demarcations.

More than three decades after McNeill's presidential address and Novick's landmark study, a clear need persists for common ground on which to articulate a more rigorous, reflective epistemology and historiographic balance. Too many historians still rely on old ideals of historical objectivity and professional demarcation. Exploring interdisciplinary approaches to understanding different narrative forms and the relations between myth and history seems essential. Surprisingly, we begin this exploration with important insights from an unlikely source, the celebrated physicist Richard P. Feynman.

As McNeill addressed the American Historical Association in New York, Feynman's book *QED* was being delivered to bookstores across Manhattan. In his bestselling popularization of quantum electrodynamics, Feynman introduces a brief note on storytelling. After three pages of grand narrative on the history of physics, from the scientific revolution to the present day, he abruptly interjects the following:

> By the way, what I have just outlined is what I call a "physicist's history of physics," which is never correct. What I am telling you is a sort of conventionalized myth-story that the physicists tell to their students, and those students tell to their students, and is not necessarily related to the actual historical development, which I do not really know![65]

At first glance, it seems surprising that Feynman is so self-aware and honest about using this "physicist's history" or "myth-story"

in place of some well-researched history produced by professional historians of science. Feynman makes no attempt to reflect
further on his surprising interjection. Instead, he leaves it hang-
ing and boldly continues with his myth-story without contextualizing his comments. Fortunately, Feynman's prolific writing
and interviews throughout his career shed light on his interjection about myth-stories.

A short 1973 documentary, "Take the World from Another
Point of View," probes Feynman's attitude toward history more
deeply. Windswept, in loafers, holding tightly to his young
son's hand, Feynman carefully navigates a wet cobblestone
slope as a voice-over introduces the Nobel laureate to a local
television audience in Yorkshire, England. As the interview
begins, the film's gaze shifts from Feynman the vacationing
family man to Dr. Feynman, the Nobel Prize–winning physicist, explaining how he interrogates the book of nature. The
audience now sees a solitary scientific genius strolling through
a tranquil Yorkshire landscape, a man with complete confidence in his intellectual pursuits. The passive natural setting
starkly contrasts with Feynman's feverish passion. Feynman
reflects on how he plies his craft as a theoretical physicist and
then communicates, to a largely nonscientific audience, an
ideal caricature of science.[66]

At the time, Feynman (see Figure 1.2) was already one of the
most accomplished physicists in the international physics community, well on his way to becoming a celebrity to the general
public. His fame would rise so high in the 1980s that the name
"Feynman" became synonymous with genius.[67] While Einstein
was undoubtedly the first modern celebrity physicist, Feynman
was his heir apparent. Like Einstein, Feynman approached science
in a fashion that contemporaries characterized as uncompromis-
ingly devoted to originality and skepticism.[68] Throughout his
career, he staunchly advocated ideals of science perhaps best represented by an unwavering commitment to an idealized picture
of scientific practice.[69]

Figure 1.2   Richard Feynman reclining in a garden circa 1984. (Source: Photograph by Tamiko Thiel. Courtesy of Tamiko Thiel.)

These scientific ideals were embedded in Feynman's worldview early in his childhood. During the summer of 1933, a fifteen-year-old Feynman busy teaching himself analytic geometry and calculus set out from his home in Queens, New York, with his family, heading west on a road trip to the World's Fair in Chicago.[70] The fair celebrated "A Century of Progress," with its ubiquitous motto "Science Finds, Industry Applies, Man Adapts."[71] All this significantly affected the young Feynman. One can imagine the wide-eyed fifteen-year-old taking the exhilarating Sky Ride, suspended on cables between two six-hundred-foot towers, then making his way to the Hall of Science, an enormous structure dedicated to instilling unbounded hopefulness in scientific and technological progress.[72] Engraved on a plaque within this great hall was the 151-word "concentrated history" of science penned by the Hall of Science's curator and retired historian of science, Henry Crew.

Apart from setting up the hall with exhibits and models that showcased scientific progress over the past century, Crew had produced his historical "masterpiece of succinctness" that alludes to an inevitable march toward an objective understanding of the universe, inviting young readers to join the march and make their mark. It begins with the lines "Pythagoras named the Cosmos; Euclid shaped Geometry; Archimedes Physics," and leaves the future of science open and inviting, ending with the promise that "Planck's quantum and Einstein's relativity theory open new epochs of science."[73] While today we might refer to Crew's brief narrative as a myth-historical masterpiece of Whiggish history rather than succinctness, it was not intended to be historically inclusive. Its rhetorical aim was to inspire visitors to the 1933 World's Fair to consider science an indispensable pillar of modern life. The argument was that a commitment to science is a commitment to progress. The implication was, perhaps, that without science, no progress exists. The scientific ideals that Feynman gleaned from his father and teachers were certainly reinforced by his visit to the Hall of Science in Chicago during the summer of 1933.

Eight minutes into the Yorkshire documentary film, the interviewer asks Feynman about his take on scientific innovation. The physicist responds: "A thing like the history of the idea is an accident of how things actually happen and if I want to turn history around to get a new way of looking at it, it doesn't make any difference.... The real test in physics is experiment, and history is fundamentally irrelevant."[74] This might seem an inflammatory and dismissive characterization of history, but what does Feynman really mean by irrelevant? He doesn't claim that history is absolutely irrelevant. Instead, he is making the important point that, in science, the history of an idea is less important to a physicist than whether experimental and theoretical evidence supports that idea. The implication? If a physicist needs to twist and turn history to make a scientific argument, so be it.

This was not Feynman's first reflection on the history of ideas and storytelling. His 1965 Nobel lecture begins by telling his audience that scientists tend to publish articles in scientific journals, framing their work as finished and polished. They "cover [up] all the tracks" and omit any mention of their blind alleys and wrong ideas.[75] Feynman was honest and unapologetic about this historical pruning. He then doubled down, determined to tell an entertaining story regardless of its historical veracity:

> So, what I would like to tell you about today are the sequence of events, really the sequence of ideas, which occurred, and by which I finally came out the other end with an unsolved problem for which I ultimately received a prize. . . . I shall include details of anecdotes which are of no value either scientifically, nor for understanding the development of ideas. They are included only to make the lecture more entertaining.[76]

For Feynman, engaging with his audience to get a rise out of them in an entertaining way was paramount; it superseded any commitment to historical veracity. Feynman was a master storyteller and generally seemed very comfortable curating historical anecdotes and employing myth-stories, but he never articulated any deeper reflection on the assumptions inherent in their use or the impact that these myth-stories might have on audiences. However, eight years after the publication of *QED*, another physicist and Nobel laureate, Leon Lederman (see Figure 1.3), adopted Feynman's language and refined the concept of myth-stories into myth-histories. In doing this, Lederman gives us a surprisingly thoughtful reflection and justification for scientist-storytellers' intentional use of myth-histories.

Leon Lederman's Myth-Historical Filters

In his 1993 book, *The God Particle,* written with science writer Dick Teresi, Lederman presents a four-hundred-page sweeping narrative covering what he initially describes as a "history of particle physics" made up of a "string of infinitely sweet moments that scientists have had over the past 2,500 years."[77] In an introductory chapter, he refers to scientific innovation as a single road that begins in Miletus in 650 BC and will end at the gates of a future mythical city "where all is understood—where the sanitation workers and even the mayor know how the universe works."[78] Lederman's cheerful

Figure 1.3 Leon Lederman standing in front of Robert Wilson's "The Mobius Strip" at Fermilab, circa 1988. The occasion commemorated the announcement of his Nobel Prize in Physics. (Source: Photograph by Reidar Hahn, courtesy of Fermilab History and Archives Project, Fermilab.)

optimism and entertaining story take readers on an epic journey along a road of inevitable progress, pausing only to admire the larger-than-life monuments raised to scientific heroes, including Democritus, Archimedes, Copernicus, Galileo, Newton, Einstein, and Feynman. With many colorful and humorous anecdotes, *The God Particle* was intended to educate a general audience on the ideals of scientific practice and to argue for the central importance of science to modern society. It is an engaging tale, and Lederman does an admirable job of leveraging historical anecdote with political pragmatism and an idealized philosophy of science.[79]

Toward the end, Lederman abruptly states, "Heretofore, as we have seen in our myth-history, ... "[80] Wait a minute! Hold your horses! What exactly is a myth-history, and why are we just now being told that the road we have been traveling for almost four hundred pages is not the history we were sold at the outset? Lederman does not bother to contextualize or explain his use of the term myth-history here; however, if the reader persists past his Acknowledgments, there is a brief yet revelatory postscript titled "A Note on History and Sources."[81] Here, Lederman reveals his approach to history and storytelling.

The author begins by admitting that "When scientists talk about history, one must be alert. It isn't history as a professional, scholarly historian of science would write it. One could call it 'fake history.' The physicist Richard Feynman called it conventionalized myth-history."[82] While this admission is refreshingly self-aware and transparent, it is also surprising that we only learn this in a buried postscript, after consuming a long narrative celebrating the so-called history of a scientific road to progress, written in a highly authoritative voice with no hint of historical contingency. Not content merely to reveal that his story fails as a proper historical account, Lederman justifies using "myth-history" as a pedagogical tool. He readily acknowledges that the "string of infinitely sweet moments" in his narrative simplifies a process that is far more complicated in real life. "The evolution of a new concept in science can be enormously complicated.... Historians sort all of this out and create a vast and rich literature about the people and concepts."[83] So far in his postscript, Lederman seems forthright in admitting that his narrative is not a proper historical account; he even gives credit to professional historians who toil to produce rich and faithful, scholarly, historical analyses. Yet his pedagogical justification for using myth-histories is still not clear.

As someone trained as both physicist and historian, I'm vexed by the next sentence, which seems critical to Lederman's justification: "However, from the point of view of storytelling, myth-history has the great virtue of filtering out the noise of real life."[84] Herein lies all the tension in the hyphen conjoining "myth" and "history." It is also critical to unpacking Einstein's black box. As a scientist, Lederman must filter out unwanted distortive noise in order to find and isolate the desired signal. In experimental lab sciences, much painstaking work is dedicated to actively cleaning and isolating signals from the surrounding noise so that we might detect and collect data that can help formulate verifiable knowledge about the systems we study (see Figure 1.4). The "signal" represents the particular data the scientist seeks from the experimental setup, while "noise" is considered any distortive artifact that arises while running the experiment. Everyone can relate to the frustrating struggle of trying to communicate by cell phone when you are far from a cell tower and the signal is faint. Your friend's voice is chopped up, interrupted, badly distorted. It's similar to hearing your favorite radio station finally consumed by the hiss of static as you head off on a road trip. In these situations, noise has overwhelmed a faint signal, depriving you of clear communication.

Signal to Noise

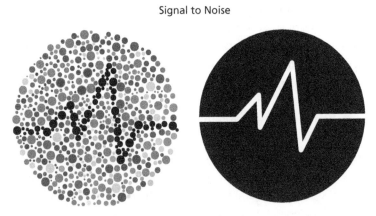

Figure 1.4 In the juxtaposition of these images one can see how filtering out the noisy image on the left can leave us with a clean signal on the right. In physics, we might say the image on the left has a low signal-to-noise ratio, whereas the image on the right has a high signal-to-noise ratio. (Source: Illustration by Zeyu [Margaret] Liu.)

What can we do to clean up this communication and restore our signal? Well, if one has access to the source of the signal, it can be selectively amplified so that it overcomes the noisy context around the receiver. Also, special filters can be installed to lower the level of undesired noise traveling with the signal, thereby cleaning the signal and making it easier to reconstruct. Either approach can increase the signal-to-noise ratio, making communication more effective. As a scientist-storyteller, Lederman believes that using myth-history is justified because it becomes an effective filter of historical noise. As a scientist, one can completely understand his instinct to increase the signal-to-noise ratio. But it is not clear that Lederman has done enough to explain what constitutes historical "signal" and what can be considered distortive "noise." Depending on perspective, one person's noise might be another person's signal. Thus, while intending to reduce historical noise, applying a myth-historical filter may be interpreted as an intervention that filters out critical information about the practice and evolution of science. This may be seen as distorting the historical signal.

In his justification for crafting a myth-history, Lederman boldly identifies historical noise as none other than the details of "real life." As a historian, one might find this idea immediately problematic. How can you filter out real life and still reliably understand scientists, their practice, and the particular historical context in which they lived? For a historian, the answer is self-evident. You can't. Also, if real life is your noise, what is your signal? To his credit, Lederman addresses that point explicitly. In discussing the historical sources of his narrative, he states: "There may, in fact, be no source for some of the best stories in science, but they have become such a part of the collective consciousness of scientists that they are 'true,' whether or not they ever happened."[85] Apparently, for Lederman, anecdotes—true or not—are the historical signals his narrative deploys. He claims that even myth-historical stories that never happened would still be "true," to some extent, because they live in scientists' collective consciousness. Later in this chapter, we will situate Lederman's claim within a broader scholarship regarding mythology, history, truth, and the role of historical anecdotes, but for now, let us take Lederman at his word.[86]

Although it is not immediately clear what Lederman means by "collective consciousness" and "true" in this context, he has briefly allowed us to peek behind the veil of his intent.[87] One thing Lederman makes clear is that when he uses the term "true" in reference to his myth-history, he is not referring to an objective historical truth based on factual details. Unfortunately, this somewhat muddies his claim to truth. The question remains: if Lederman is not trying to teach historical facts, what truth does he want to convey by using myth-history? No matter the answer, it can no longer be denied that myth-histories of science are not simply a sloppy copy of scholarly history. As we saw in the case of Weinberg's approach to history, myth-history is a fundamentally different approach to studying the past. It is a distinct historical lens.

This myth-historical lens is not at all passive. It is used as an active filter; selecting and filtering out historical details that may be incongruous with contemporary and collective scientific imaginaries. These discarded details usually correspond to Einstein's socially conditioned science in the making. In addition to filtering, these myth-historical lenses infuse historical reconstructions with curated, anecdotal stories that project ideal mythological tropes of scientific practice. What emerges from this process works to reinforce existing idealizations of science. These tend to look very much like Einstein's objective and complete characterization of science, a heroic climb toward inevitable progress (see Figure 1.5).

Do these myth-historical reconstructions represent a deeper essential truth about science, or are they simply wishful distortions? Do we need these myth-histories to better understand the unique character of scientific progress? What are the effects of using myth-histories? These are the questions we will grapple with in the four case studies in this book, but to begin to answer them, we need to track Lederman's intent further—and to query his assumptions about what constitutes scientific practice and its evolution.

Myth-Historical Filter

Messy Science in the Making   Objective Idealized Science

Figure 1.5 Lederman's concept of myth-history can be understood as a historical lens distinct from scholarly histories. The lens filters out the "details of real life"—science in the making—that are inconvenient or incongruous with collective scientific imaginaries. What emerges from this process of filtration is a projection of science, infused with scientific ideals that highlight a heroic and inevitable rise up the escalator of progress. (Source: Illustration by Zeyu [Margaret] Liu.)

As we learn from Lederman's Preface to the second edition of *The God Particle*, published in 2006, the book was written primarily to help lobby Congress to ensure continued funding of the massive superconducting super collider (SSC) particle accelerator. In the early 1990s, the SSC was being built in Waxahachie, Texas.[88] Completed, it would have been the largest, most powerful particle accelerator ever built, capable of probing the depths of matter and the earliest moments after the Big Bang. According to Lederman and other champions of the SSC, including Steven Weinberg, such a large accelerator was necessary for finding the Higgs boson (Lederman's "God particle"), completing the standard model of particle physics, and possibly uncovering an underlying and fundamental "theory of everything."[89]

Although the SSC had been in the works since 1983, and nearly $2 billion had already been spent, the U.S. Congress shuttered the program for good in the fall of 1993. Much had changed in the interim. The end of the Cold War had suddenly shifted geopolitical considerations and domestic spending priorities. For decades, strategic alignments between "big science" projects like the SSC and national defense had led to almost unquestioned political and economic support. As the American political landscape settled into a new post–Cold War order, the SSC found itself "dissociated from national security" and extremely vulnerable to domestic politics.[90] With continuing cost overruns and a rising price tag that had ballooned to $11 billion, long-standing critics of the SSC won the day. The fascinating backstory of how the SSC found itself on the congressional chopping block is a wonderful example of the human and social forces that undergird all science.[91] Although these are details generally filtered out of myth-historical tales of progress, they add important contextual clues about how science navigates its social constraints. Preserving and spotlighting these details could actually benefit science. Careful analyses of case studies like the shuttering of the SCC project could suggest new and better strategies for scientists—and nonscientists—to manage contentious issues of science funding.

Within the context of fierce congressional battles over appropriations, Lederman became a champion of the fledgling SSC. At the time, Lederman was a recent Nobel laureate and director of the Fermilab particle accelerator. As a high-profile physicist with appropriate scientific expertise, a penchant for humor, and a clear agenda, he seemed like an ideal choice to coauthor an entertaining, accessible book that could build good will while informing a general audience about the importance of science, especially particle physics. More important, with the future of the SSC in grave danger, he was unapologetically attempting to build political capital, lobbying Congress for continued funding.

Lederman's Historiographic Interlude

In Lederman's myth-historical narrative there is a telling Interlude titled "The Dancing Moo-Shu Masters."[92] A closer look at this Interlude suggests it undergirds much of the rationale for his use of myth-history. Lederman appears concerned that some readers will not pick up his implicit descriptions of what science is and how it progresses as illustrated by his myth-historical progress narrative. So he pauses his entertaining heroic story for what can only be described as a cautionary public service announcement that warns of an existential threat to science: the lay public's misunderstanding of what science is and how it "evolves and progresses."[93]

Lederman begins by describing an exchange with Louisiana Senator Bennett Johnston; he is disturbed to find that a "U.S. Senator, hungry for knowledge" about physics, had looked to Gary Zukav's book *The Dancing Wu Li Masters*.[94] He targets Zukav's book and others like it, because, he believes, while the authors pretend to introduce their audience to the latest ideas in physics, they tend to "jump from proven, solid concepts in science to concepts that are outside of physics and to which the logical bridge is extremely shaky or nonexistent."[95] To Lederman, this is a hallmark of pseudoscience, and he bristles at the thought that Zukav and other more extreme "charlatans" might misrepresent and distort scientific practice, corroding public faith in science. Although Lederman seems to make delineations between science and pseudoscience with clear and unwavering authority, readers trained in history of science, STS, or philosophy of science know that this "demarcation problem" has been a canonical thorn for decades.[96] In particular, David Kaiser's incisive work in *How the Hippies Saved Physics* clearly problematizes Lederman's demarcation.[97]

Although Lederman does not name it as such, his Interlude is a clear example of a scientist engaging in "boundary work."[98] For decades, scholars have tried to clarify the criteria that make scientific practice unique. Each time, the demarcation between science and pseudoscience has resisted rigorous definition. If the demarcation criteria are too easily met, the bar is set too low and we end up including activities that most scientists would not consider scientific. On the other hand, if the criteria are defined too rigidly, the bar will be set too high, thereby excluding activities considered scientific by most practicing scientists. This is clearly an important problem for anyone discussing science. Defining this human activity would seem like a good place to start when trying for an intelligible conversation about it. Myriad notions of scientific methods or ethical behavioral norms have been proposed as litmus tests or criteria for what constitutes scientific practice, but every time the demarcation problem has persisted.

In his Interlude, Lederman avoids any mention of this rich discourse, yet attempts to sketch out his own solution to the demarcation problem. He notes that, like all disciplines and social "fields of endeavor," science has an establishment.[99] Yet one of the distinguishing characteristics of science is that "iconoclasts, rebels with (intellectual) bombs, are sought after zealously—even by the science establishment itself.... This [sacred obligation]—that we should remain open to the young, the unorthodox, and the rebellious—creates an opening for the charlatans and the misguided, who can prey upon scientifically illiterate and careless journalists, editors, and other gatekeepers of the media."[100]

Lederman briefly acknowledges the social reality of a rooted establishment, but considers science's hallmark to be its commitment to iconoclasts and rebellious youth. This commitment is so important, he even terms it a "sacred obligation."[101] Lederman's overemphasis on youth and divergence as a hallmark of science is exactly the myopic picture Thomas Kuhn challenged in his groundbreaking work sixty years ago. What about convergence? Where is Kuhn's "essential tension"? Where is the constant pull between convergent normal science and divergent revolutionary science (see Figure 1.6)?[102] Although Lederman never explicitly refers to Kuhn, essential tension, paradigms, or paradigm shifts, he does call on the "patron saint" of physics, Richard Feynman, who, consciously or not, seems to somewhat echo Kuhn's notion of essential tension: "Each generation that discovers something from its experience must pass that on, but it must pass that on with a delicate balance of respect and disrespect."[103]

Figure 1.6 Kuhn's essential tension. In a 1959 essay, Kuhn sketched out preliminary ideas that would evolve into the underlying foundation for his *Structure of Scientific Revolutions* (1962). The communal behavior of scientists shows an essential tension at the heart of science between convergent and divergent forces. These opposing forces are both necessary for our understanding of scientific progress. (Source: Illustration by Zeyu [Margaret] Liu.)

Regardless of his knowledge of Kuhn and the demarcation problem, Lederman considers science's sacred commitment to open-mindedness and divergence to be a double-edged sword. It is critical to scientific progress and simultaneously at the root of its susceptibility to pseudoscientific "charlatans" who can deeply influence the naïve and scientifically illiterate public. By this reasoning, Lederman considers reckless pseudoscientific infiltration of science as a severe danger to the greater public good and something to actively guard against. Like a knight on a crusade, Lederman seems poised to slay pseudoscience, ensure the purity of science, and defend the realm at all costs (see Figure 1.7). His "Dancing Moo-Shu Masters" interlude is a critical cog within his larger argument highlighting the exceptional nature of science, how it progresses, and its importance to modern society.

For Lederman and many practicing scientists, the advancement of science proceeds in what he calls successive revolutions "executed conservatively and cost-effectively."[104] He is careful to explain that these successive, divergent episodes do not vanquish and replace older scientific worldviews. Instead they retain established knowledge and extend it to new domains. Each revolution

Figure 1.7 "Sir" Leon Lederman, Champion of Scientific Myth-History, Defender of the Realm. Pictured here at a Fermilab-wide party. (Source: Courtesy of Fermilab History and Archives Project, Fermilab.)

reveals more and more about the workings of the universe. In order to make this conceptualization of cumulative scientific progress more accessible and concrete, Lederman uses pictures of concentric, embedded, ellipse-shaped knowledge domains to help explain his ideas. The first domain begins with Archimedes's work on statics and hydrostatics from 100 BC. The subsequent scientific revolution in Lederman's "march of progress" is attributed to Galileo's work in 1600 CE, in which he engulfed and extended Archimedes's ideas into novel domains like kinematics (balls rolling down inclines) and celestial dynamics (motions of the moons of Jupiter). The Galilean revolution was followed by successive revolutionary extensions attributed to Newton, Maxwell, Einstein, and, more generally, to "quantum physics" (see Figure 1.8).[105]

Interestingly, Lederman recognizes that these revolutionary episodes produce a "good deal of waste as well."[106] In this way, he

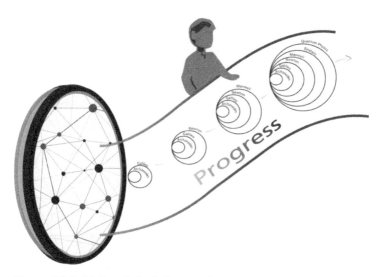

Figure 1.8 In his Interlude, Lederman depicts progress using concentric ellipses that push the envelope of knowledge. Here his ideas of concentric ellipses are superimposed on the progress escalator emanating from the myth-historical lens (see Figure 1.5). (Source: Illustration by Zeyu [Margaret] Liu.)

distinguishes between past ideas, like Archimedes's hydrostatics, that are right, and most ideas, theories, and suggestions that are wrong. "In the contest for control of the frontier there is, in terms of concepts, only one winner. The losers vanish into the debris of history's footnotes."[107] This statement is revealing. Lederman seems to describe scientific innovation as an objective, mechanistic, purely rational process that decides whether a novel scientific concept is right or wrong, winner or loser. As such, at the heart of Einstein's black box is a purely rational process of discovery represented by idealized tropes like the much-celebrated scientific method. Does scientific innovation really work that way?

For many, a commitment to a universal scientific method should ensure that it does, but as scholars of science have shown, much of what is considered scientific research does not necessarily subscribe to this methodological ideal. Ultimately, facts and data do not speak for themselves. Scientists must always interpret their data using predetermined scientific theories and models that are products, tools, and artifacts of particular historical contexts. Scientists do not just passively read some archetypal "book of nature"; they are continuously, actively interpreting a mediated reality. Sayings such as "reading the book of nature" are powerful metaphors, but they fail to accurately represent the daily grind of scientific practice. The demarcation problem persists because we cannot identify a universal common practice to define all of science. As a result, there is also no clearly agreed-upon litmus test for whether a scientific idea or result is right or wrong. The human and social dimensions of science must always be taken into account. The disappearance of conceptual losers from the progress of science does not happen naturally or passively, contrary to Lederman.

The saying "winners write history" is appropriate here. In the history of science, those representing the winning concepts and ideas forge a scientific consensus, leveraging reimagined myth-historical narratives to support and defend that consensus. To use Lederman's metaphor, this grand narrative construction actively and deliberately preserves a desired signal by filtering out unwanted noise. In reality, what is filtered out in these myth-historical constructions are not just irrelevant details, but ideas, concepts, and scientific results considered to challenge or question the winning worldview. These omissions are not Lederman's passive "vanishings" and they are certainly not historically inevitable. They are historically contingent results based on human and social activities that affect scientific practice, and they reflect a messier past filled with contentious dialectics that have been purged from myth-historical narratives. The ideas and concepts carried over from one revolutionary ellipse to the next in Lederman's pictorial representations are certified as the scientific "winners" and therefore considered correct. Anything left out is assumed to have been wrong. Far too many historical examples of vanquished, filtered, and forgotten ideas that reemerge decades or centuries later prevent us from taking at face value this seemingly inevitable image of scientific advancement.

As a result of all Lederman's justifications and explanations in his interlude and postscript, the term myth-history seems like a remarkably appropriate characterization of this type of Whig narrative. He clearly recognizes that his historical reconstruction falls short of veracity and scholarly rigor, yet that was never the intent of his book. So, what truth does this myth-history convey? What seems most important to Lederman is to teach the archetypes of science, reinforce the ideals bound up in prevailing scientific social imaginaries, protect scientific exceptionalism, and simultaneously to rationalize the current scientific worldview.

Myth-histories like Lederman's are powerful social agents that work to advance these causes. They are employed by scientist-storytellers as pedagogical and rhetorical tools. However, it is imperative that we now go beyond their insider meanings and intents and explore the ideological foundations and impacts of myth-history as a historiographic category from the perspective of historians and other scholars of science. In the following section, we further unpack what Lederman, Feynman, Weinberg, and scores of other scientist-storytellers are doing when they consciously, or unconsciously, employ myth-history.

Myth-History as Chimeric Narrative Category

As McNeill noted in his 1985 presidential address to the American Historical Association, myth and history are not polar opposites. Myths should not be associated with falsehoods, or history with some absolute or objective Truth. Myths and histories are focused on two different versions of truth. As such, myth-history should be understood as a fundamentally "chimeric" category.[108] The hyphen joining the two narrative modes is critical. Myth-historical reconstructions are not purely fantastic tales; instead, they do the work of mythologies by infusing elements of collective imaginaries, while remaining tethered to real historical chronologies (see Figure 1.9). They are a different species of narrative that manages to retain aspects of both myth and history. In doing so, myth-histories refract reality in particular ways, revealing truths about science that fundamentally differ from historical truth. They are socially situated and culturally encoded; ultimately they exert wide-ranging effects in ways scholarly histories do not. With this in mind, we can begin articulating a framework for McNeill's new epistemology, which seeks "historiographical balance between Truth, truths, and myth."[109]

Since scientific myth-histories are a chimeric construction and retain key characteristics of both mythological and historical narratives, it is worth taking a moment to consider more concretely how these two forms intermingle. This is no easy task; cross-pollination is messy. Myths alone can be considered "sophisticated social representation[s]" that result in "a complex relationship between history, reality, culture, imagination and identity."[110] Yet if we are to have any hope of understanding how myth-history could be used to help transform Einstein's science in the making into a science that is complete and objective, we must take the

Myth-Histoy: Infused Historical Scaffolding

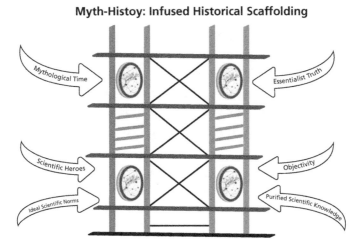

Figure 1.9 Myth-history is a chimeric historiographic category. It uses a historical scaffolding and infuses it with mythological tropes such as scientific heroes, objectivity, ideal scientific norms, and purified scientific knowledge. These tropes are informed by, and reinforce, collective imaginaries. (Source: Illustration by Zeyu [Margaret] Liu.)

plunge. While certainly not an exhaustive interrogation of this chimeric category, the following discussion will help identify important features and begin to outline its form.

All myths are tied to a particular time, place, and cultural setting, yet to some scholars there is a timelessness that makes myths' social function recognizable no matter what the context. Since the nineteenth century, anthropologists, cultural historians, sociologists, and other scholars have done important ethnographic and theoretical work on the origin, function, content, structure, and meaning of mythology.[111] Even with the broad recognizability of myths and all this scholarly work, there is still no clear set of criteria that fully describes myth as a narrative category.[112] Instead of attempting to produce an exhaustive list of criteria for myth-history, the following section explores four themes that seem indispensable to a common-ground understanding of the chimeric tensions within scientific myth-histories.

Myth-Historical Time

In myth-histories, as in Whig histories, the present has a privileged perch from which the past is reinterpreted and the future is imagined. Scientist-storytellers rely on historical timelines as scaffoldings, but as their stories unfold there is a sense in which they jump in and out of chronological linear time to explore what anthropologist Claude Lévi-Strauss called "mythological time":

> On the one hand, a myth always refers to events alleged to have taken place in time: before the world was created, or during its first stages—anyway, long ago. But what gives the myth an operative value is that the specific pattern described is everlasting; it explains the present and the past as well as the future.[113]

Lévi-Strauss's notion of mythological time as distinct from linear historical time is critical to our understanding of how scientists use myth-histories. This fungible use of time as both linear and nonlinear allows scientist-storytellers to achieve several rhetorical goals simultaneously. First, by grounding their narratives in established historical timelines, they can camouflage the chimerism of their narratives and present their myth-histories as authoritative histories for popular audiences. As a result, most lay audiences make no clear distinction between history and myth-history. Except to professional historians and lay people well-versed in the history of science, most of these myth-histories seem like authoritative history that is rigorous yet accessible.

It is rare to find scientists as forthright as Feynman about the historical shortcomings of their narratives. Even Lederman, who gave us a peek behind his myth-historical veil of intent, did so only in a buried postscript. Most scientist-storytellers, like their audiences, remain unaware that the history they recount differs fundamentally from scholarly history. The inherent tension in their chimeric narratives is hidden and latent. However, when confronted by historians about the inaccuracies their narratives propagate, some, like Weinberg, defend their scientific myth-histories by pointing to the exceptional nature of science. It is a human activity fundamentally different from art or politics. The cumulative, rational foundations of science justify a Whiggish framing. Any inaccuracies due to myth-histories are minor inconveniences on the journey toward a greater good. For these scientist-storytellers, historical veracity is "fundamentally irrelevant"; what really matters is the search for scientific truth.[114]

To that end, the second important rhetorical effect that arises from the use of mythological time is a rational reconstruction of the scientific past. This is an explicit transformation of what actually happened in history into what Pierre Hohenberg noted "should have happened."[115] This use of mythological time to fill historical cracks and correct the past is akin to what some anthropologists observe in their fieldwork. For example, Jonathan Hill notes that "South American mythic histories attempt to reconcile a view of 'what actually happened' with an understanding of what ought to have happened."[116] Yet what "ought to have happened" depends on the particulars of perspective and context. It depends on who holds power in the present.

So these myth-historical explorations are intimately related to questions of power structures within current scientific communities. The present serves as a pivot from which these communities work to control the past so that they can more effectively "redefine the future."[117] The emphasis in these narratives is to establish coherence in scientific progress, reframing the latest scientific understanding as the momentary pinnacle in an inevitable, natural process of discovery.

Last, employing mythological time allows the audience to momentarily escape from the present, entering a dimension of timelessness and universal translation. In this realm, scientific past, present, and future commingle and merge (see Figure 1.10). As a result, anyone can theoretically identify with a scientific hero born centuries ago in some faraway land. This fluidity buttresses the Mertonian norm of scientific universalism, insisting that science has no allegiance to a specific culture or context.[118] It is a rational enterprise, which assumes that all humans are rational agents who,

Figure 1.10 Myth-historical time uses historical chronology while also exhibiting the essence of universal time in which past, present, and future seem to merge. (Source: Illustration by Zeyu [Margaret] Liu.)

given correct methodology, can engage the natural world from any cultural and historical context and arrive at universal truths.[119]

Myth-Historical Facts: Fungible Fact Hierarchies

Scientific myth-histories are complex collages of historical fact and fiction. They are pilgrim tales of past heroes on epic journeys of scientific discovery. They are links in a chain of scientific progress welded together by hyperbole and caricature. As Lederman noted, the links themselves are made from some of the "best stories in science," and "are 'true,' whether or not they ever happened."[120] Again, this playful use of truth by Lederman may shock professional historians of science, who value the veracity of historical facts above all else; but for a scientist trying to use myth-history for particular rhetorical purposes, historical facts and historical truth are easily fungible.

In his introductory chapter to *The God Particle*, Lederman refers to one of the core tenets of scientific practice as rational argumentation that "rigorously exclud[es] superstition, myth, and the intervention of gods."[121] In this context, he groups mythology with superstition and religion, then rigorously excludes them as distortive noise. Apparently, when it comes to practicing science, experimentation and a rational exploration of objective reality seem to be signals, while mythology, superstition, and religion are considered merely sources of noise.

Yet in reconstructing his myth-history, Lederman explicitly inverts this relationship and labels myth-historical stories as somehow "true" and real-life experiences as noise. Such brazen fungibility may seem hypocritical, but from the perspective of many scientist-storytellers, it makes perfect sense. It also reveals an underlying bias around a particular hierarchy of facts. It's the same fact hierarchy that Feynman invoked by claiming that "the real test in physics is experiment, and history is fundamentally irrelevant." All facts are not created equal.

For scientist-storytellers, facts about the natural world and facts about the essence of scientific practice are significantly more important than concrete historical facts. Therefore, confronted by a conflict between these, a scientist's professional bias and integrity will invariably allow for the tweaking of history to accommodate natural facts or facts about what they consider to be the essence of science. However, from the point of view of a professional historian of science, whether natural facts are right or wrong is always secondary to the historical facts that elucidate the specific context in which they were presented. These different fact hierarchies are critical to grasping the boundary tensions between myth-histories and their narrative cousins in the history of science.

There is a telling parallel between the long-standing demarcation problem in distinguishing science from pseudoscience, and the demarcation problem that distinguishes scholarly history from myth-history. Both demarcation problems seem to demand the establishment of a clear boundary between practices that are acceptable to a community and practices that are not. However, if we reflect on how fact hierarchies affect demarcation problems,

we quickly realize that a pivot from epistemological fundamen-
talism to epistemological pluralism may help lead to a more tol-
erant, accepting, and reflective discourse.

Myth-Historical Imaginaries: Truths Greater than Truth Itself

Myth-histories share similarities with another genre of narrative
fiction, magic realism, made famous by Colombian author and
Nobel laureate Gabriel García Márquez. Similar to scientist-
storytellers, García Márquez "re-writes Colombian history, inter-
weaving legendary and mythic images with historical fact, thus
sharpening the truth he wishes to convey."[122] For many
Colombians, the truth conveyed in these narratives, factual or
not, represents "the true meaning of Colombianness."[123] This is
similar to Lederman's claim that scientific myth-histories reveal
some kind of essentialist truth about science regardless of the
veracity of the anecdotal stories they tell. But again, it raises the
question, How can we better understand the truth these narratives are revealing?

In his own context as a scientist-storyteller, Lederman points to
but does not examine the source material for these truths about
science. Instead of a systematic and rigorous examination of historical archives, he unveils them to be the "collective consciousness of scientists."[124] Although Lederman does not tell us explicitly
what he means by "collective consciousness," one possibility is to
draw a clear line between Lederman's repository of myth-historical
stories and the Jungian notion of "archetypes," explained as "products of the collective unconsciousness."[125] According to Jung,
these symbols or underlying themes can arise in the shared unconscious of a collective cultural group.[126] For some myth scholars,
Jungian psychoanalysis is important to understanding how
mythologies both reflect and fulfill deep-seated social needs.
Myths help us wrangle with what Lévi-Strauss called society's
"inherent possibilities and its latent potentialities."[127]

Yet Lederman seems to refer to stories that inhabit a collective
consciousness, not a collective *un*consciousness. As a result, his

claim can also be understood in conversation with the evolving study of collective or cultural memory.[128] This rich multidisciplinary framing involves exploring the "interaction between the psyche, consciousness, society, and culture."[129] Over the past several decades, the early work of Émile Durkheim and Maurice Halbwachs on collective representations and memory have been rethought and extended in fascinating ways, resulting in pictures of cultural inertia that transcend static notions of tradition and reflect a more complex sociocultural dynamic.[130] The discourse, situated in the interpenetrating realms of the collective conscious and unconscious, allows us to more carefully examine the underpinnings of Lederman's claim to truth.

This idea of a dynamic collective memory dovetails nicely with the powerful and evolving framework of "social imaginaries." Imagination operates at both individual and intersubjective levels.[131] When imagination becomes part of collective social processes, then it can unite members of a community in shared perceptions of communal identity and existence. Broadly speaking, social imaginaries have been used by scholars for decades to denote "collective beliefs" about how social groups function.[132] In his work on *Imagined Communities*, Benedict Anderson found that nationalism could only be understood as a cultural artifact born out of a "complex 'crossing' of discrete historical forces" that managed to become modular, mobile, and mutable.[133] As such, these artifacts have evolved to work on imagined political communities by ensuring that imprinted on each individual mind "lives the image of their communion."[134] For Anderson, this imagined collection of minds is "tied together through shared practices of narrating, recollecting, and forgetting."[135] This cultural artifact goes far deeper than transient political ideology; these imaginaries tend to "command profound emotional legitimacy."[136]

In a more recent incarnation of social imaginaries applied to science and technology, Sheila Jasanoff refines Anderson's notion of imaginary. She questions the lack of scholarly work on scientific and technological imaginaries, extending the concept of sociocultural artifacts to include not just communal mental states but also the collective materiality so important to studies of technoscience. For Jasanoff, "sociotechnical imaginaries" are "collectively held, institutionally stabilized, and publicly performed" understandings of social life and order.[137] These are all important qualifiers and extensions on Anderson's concept. In thinking about Lederman's discussion of myth-historical truth and the collective consciousness of scientists, it seems legitimate to think about scientist-storytellers tapping into those collectively held, institutionally stabilized, and publicly performed social imaginaries of idealized scientific practice. As such, the telling and retelling of scientific myth-histories can be understood as a ritualistic performance of sociotechnical imaginaries.

The stories of the past that Lederman and other scientist-storytellers mine to construct their myth-historical narratives are part of this broader discourse on collective memory and social imaginaries. These imaginaries are a "crucial reservoir of power and action" that needs to be studied.[138] Intentionally or not, scientist-storytellers operationalize myth-histories as rhetorical agents that help to create social cohesion, a clear sense of "we," and, by extension, a sense of "other." Myth-histories are both products and generators of social and individual identity, but all this work does not come free. Their ability to demarcate and determine truth requires tremendous power, what storyteller Chimamanda Ngozi Adichie calls the "principle of nkali."[139] Studying the operative value of these myth-histories helps us to understand the tacit imaginaries embedded within and to see the "topographies of power" hiding behind the rhetorical veil of universalism.[140]

As modular, mutable imaginaries, myth-histories are products of present-day social and cultural contexts, and thus not strictly timeless and universal. Yet they have the mythological feeling of an everlasting sense of permanence. As collages of collective memory and social imaginaries, myth-histories are materially encoded social actors that draw on and perpetuate particular cultural beliefs, rituals, values, and norms. We should avoid overly reductionist labeling of myth-histories as rhetorical distortions of historical truth. Framing these narratives as dynamic engagements with a collective conscious and unconscious may be more fruitful.

Based on recent work on collective memory and social imaginaries, there is little doubt that myth-histories have operative value as catalysts for scientific consensus and identity. So it is important that we study them as social actors within their particular contexts. Scientists tell stories in long-form books, in tangential asides in textbooks, as entertaining anecdotes within a rich oral tradition, and in the everyday rituals that produce a sense of community identity. Myth-histories come in many shapes and sizes, and as we will see in the next section, it's not just what they include, but also what they exclude that matters.

Myth-Historical Filters and Erasers

As we have seen, the signal that myth-historians like Lederman covet is not made of well-documented and contextualized historical facts. Instead, the signal is made up of a patchwork of apocryphal anecdotes that seem like a timeless and essentialist portrait of science, but are encoded with norms and values meant to resonate with present-day scientific social imaginaries. As such, they provide a single coherent framework that allows for unquestioned, seamless translation between disparate contexts and a uniform, logistically consistent progress narrative that preserves a linear causal chain of evolving natural facts. Whenever historical facts complicate or make it hard to recognize the desired coherent signal, they are simply filtered out of myth-historical narratives.

Lederman noted that myth-histories are powerful pedagogical tools with the "great virtue of filtering out the noise of real life."[141] To articulate what they consider the essentialist truth of ideal science, scientist-storytellers aim for a sense of connection with their audience. Any historical and contextual details that obscure this essentialist signal or make it more difficult for a modern audience to identify with past scientific heroes are filtered out. In this way, audiences can find escapism by immersing themselves in a seemingly timeless myth-history that allows them to have fluid scientific conversations with heroes like Galileo, Newton, and Einstein. All barriers to translation are removed, along with any inconvenient details about the hero's historical context. The heroes, their context, and their ideas are all sanitized and recast in a modern template.

It is a testament to the power of myth-historical filtering and translation that modern audiences can grasp fundamental concepts of Newtonian physics, while failing to understand Newton's eclectic work as a natural philosopher and his relationship to his contemporaries.[142] In constructing scientific heroes like Newton and making them relatable to modern audiences, myth-histories necessarily minimize much of the historical context that would make Newton appear to be anyone but an ideal scientist. To this end, the collectively sourced stories that are myth-histories constantly negotiate between "platonic" ideals and "pragmatic" contexts, between inscription and erasure.[143]

In his study of participant histories, Richard Staley has argued that "forgetting is integral to scientific advance," and yet should not limit our understanding of scientific practice and its historical development.[144] In studying scientific storytellers and their myth-histories, what has been forgotten is just as critical as what has been remembered. Forgetting plays an integral role in both individual and collective memory. It can protect against trauma by excising parts of the past, but it can just as easily be a mechanism of oppression by marginalizing undesirable people and ideas. It can be a catalyst for scientific convergence, but it can also delay progress by suppressing divergent tendencies. Whether forgetting is a passive consequence of an imperfect collective memory or the result of an intentional rhetorical filtering by scientist-storytellers to justify consensus and protect a status quo of scientific ideals, the effects of forgetting should be studied and better understood.

Conclusion

The four chimeric themes explored in the previous section are not exhaustive, but they give a foothold from which we can move on and begin examining case studies, probing the broader effects of scientific myth-histories. These narratives are "projections on the part of the collective" by individuals who participate "in order to belong." Both the collective and individual rely on cultural capital in the form of "cultural traditions, the arsenal of symbolic forms, the 'imaginary' of myths and images, of the 'great stories,' sagas and legends, scenes and constellations that live or can be reactivated in the treasure stores of a people."[145]

Intentionally or not, these stories have significant operative value that require intricate structures and topologies of power.[146] As such, it seems impossible to discuss myth-histories without considering power. We should do more than correct them, we should study them. Since "myths that pertain to a form of life tell you quite a lot about that form of life,"[147] studying myth-histories in context can be particularly revealing. They help create scientific consensus by amplifying a preferred historical signal. They filter out and further marginalize people and ideas that don't align with the status quo. Here are important questions we need to try answering. Do these narratives protect science and its social capital? Or do they feed doubt that slowly erodes public trust in science? Regardless of individual authors' intent, we should take note of their impacts on pedagogy, public discourse, and future directions of scientific research.

From the preceding analysis, it is clear that any thoughtful exploration of the boundary between history and alternate forms of self-histories, like Whig histories and myth-histories, eliminates the necessity and even the possibility of a straight polarized disciplinary conflict. Scientists and historians approach the scientific past in very different ways. Some historians appreciate the power and value of Whig histories, while others struggle with the realization that, to some extent, they, too, produce myth-historical narratives. Some scientists may be content to leave all historical inquiry to professional historians, while others consciously deploy myth-histories for pedagogical and rhetorical purposes. Ultimately, our exploration of this messy boundary has been helpful in gaining insights into the varying intents and assumptions in all forms of historical inquiry. As we have seen, alternate historical epistemologies and scholarly historical narratives are really different species of storytelling that should not be compared one-to-one or made mutually exclusive.

Where does that leave us? First, both historians and scientists should strive to be more reflective and transparent about their own motivations and assumptions when creating history of science. Second, everyone involved in the boundary work discussed here could also be contributing to the establishment of a common ground that leaves them more open to different narrative forms and less dismissive of alternate approaches. Finally, we need to be more aware of the impact that the various forms of narrative can have on scientific communities and the public at large. The following four chapters examine case studies that help us understand the power and consequences of propagating alternate historical epistemologies like myth-histories. The goal is to look into the black boxes that transform Einstein's subjective science in the making into its objective, idyllic past.

2

Myth-Historical Quantum Erasure

The Case of the Missing Pilot Wave[1]

> Philosophy of science is just about as useful to scientists
> as ornithology is to birds.[2]
>
> —STEVEN WEINBERG

Quantum Omissions

In 1982, John S. Bell (see Figure 2.1), one of the most celebrated
physicists of the twentieth century, wrote "On the Impossible Pilot
Wave," a paper that questioned the dominance of the Copenhagen
interpretation of quantum theory.[3] His chief complaint was the
complete omission from scientific discourse and pedagogy of alternate interpretations of quantum theory like Louis de Broglie's 1927
deterministic pilot wave interpretation.[4] Bell's frustration with his
physics education was palpable: "But why then had [Max] Born not
told me of [the de Broglie] 'pilot wave'?... Why is the pilot wave
picture ignored in text books? Should it not be taught, not as the
only way, but as an antidote to the prevailing complacency?"[5]

In context, however, Bell's critique expresses more than frustration with his physics education at Queen's University Belfast
during the late 1940s.[6] His critique also assails a "prevailing complacency" within the physics community, which, according to
Bell, had too easily settled on the Copenhagen interpretation of
quantum theory as orthodoxy, while ceasing to question its foundations. At the heart of this orthodoxy lay powerful ontological
claims about the nature of reality that were at odds with classical

Figure 2.1 J.S. Bell in his office at CERN, commenting on his eponymous inequalities circa June 1982. (Source: Photograph by CERN PhotoLab. Courtesy of CERN.)

physics and everyday human experience. Bell complained that these ontological claims, such as "vagueness, subjectivity, and indeterminism," had been elevated without experimental justification.[7] They should be treated not as a necessary part of underlying physical reality but as a historically contingent and "deliberate theoretical choice."[8]

During the mid- to late 1920s, as the mathematical foundations of quantum mechanics were established, there were fluid interpretive debates about what these newly congealing models of the microscopic world implied about our understanding of reality. Classical mechanics seemed to work extremely well for macroscopic objects, like marbles and baseballs, but the study of atoms and other microscopic phenomena left physicists doubting their classical understanding of reality. By the late 1920s, the ongoing quantum revolution seemed to require fundamental breaks with classical understandings of concepts like position, momentum, particle trajectories, and determinism itself.

Classically, an object has a clearly defined position and momentum, independent of an observer's ability to measure it. Our certainty about where a particle is at a given time, and where it's headed in the future, is limited solely by the precision of our measurements. In theory, there is no limit to our precision in measuring things like position and momentum, so, given the right tools, we could become certain of the particle's future dynamics. This is classical determinism, and it is absolutely forbidden under the orthodox interpretation of quantum theory, which insists that it's not only our ability to measure that's limited, but reality itself.

The physical behavior of quantum systems is dictated by a mathematical construct called a wave function, or psi (ψ), that, unlike classical observables, seems to have no direct physical meaning. Max Born's statistical interpretation of ψ, a core part of the orthodox interpretation of quantum theory, relates this wave function to an underlying probability that is unavoidable. In other words, the universe has inherent uncertainty baked into it. This uncertainty, or indeterminism, is generally associated with Werner Heisenberg's "uncertainty principle"—another pillar of the orthodox interpretation of quantum theory to which Bell bitterly objected.[9]

In addition to ψ and this inherent indeterminism, how we choose to study a microscopic quantum system directly affects how reality manifests. When studying quantum objects like electrons experimentally, we must decide ahead of time what properties to examine. For example, if we choose to study an electron's particle behavior, it will manifest as a distinct object with definite mass and position, but if the experiment is set up to study the electron's wave behavior, then it will manifest as a diffuse wave-like phenomenon (see Figure 2.2). This observer-dependent wave-particle duality is generally associated with Niels Bohr's "principle of complementarity," and is a third pillar of the orthodox interpretation of quantum theory. In sharp contrast to the classical understanding, at the quantum level, the universe seems to be

Wave-Particle Duality

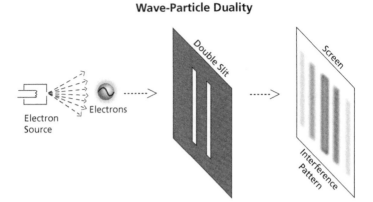

Figure 2.2 The wave or particle nature of electrons can be shown depending on the experiment chosen to study their behavior. In this case, we see that a double-slit style experiment has been set up; as a result, the electrons will show wave-like properties (an interference pattern). The electrons will only be observed hitting the screen at points that correspond to the bands and not in-between. This is not what we would expect of particles! (Source: Illustration by Zeyu [Margaret] Liu.)

fundamentally indeterministic and observer-dependent. From this counterintuitive quantum reality stem well-known quantum paradoxes, including wave-particle duality, the measurement problem, EPR (Einstein-Podolsky-Rosen), and Schrödinger's cat.[10]

Contemporary studies in the foundations of quantum theory work to understand these paradoxes by recasting them in new frameworks that make them more intelligible and consistent with our understanding of the physical universe. Over the past three decades, we have seen an explosion of interpretations of quantum theory and renewed vigor in quantum interpretation debates. Yet, in the early 1980s, when Bell complained of the "prevailing complacency," the Copenhagen interpretation of quantum theory still dominated the physics community and the broader public imagination.

Although Bell was clearly frustrated by the lack of debate surrounding interpretations of quantum theory, he was not sure how it came to be. This chapter examines the context of this long complacency. It reveals impacts of myth-historical quantum narratives that helped congeal scientific consensus around a forced orthodoxy, train generations of physicists to heed the pragmatist slogan "shut up and calculate,"[11] suspend interpretational debate, and ensure that unwelcome alternate quantum interpretations that challenged existing orthodoxy were filtered out. As Bell's frustrations seem rooted in his education, we start by examining his first encounters with quantum physics.

The Quantum (Mis)education of John S. Bell

By the time Bell wrote "On the Impossible Pilot Wave" in 1982, he had realized that his education at Queen's University Belfast was not abnormal. Thanks to a confluence of factors, including the early canonization of quantum textbooks, a rising pragmatism, and de Broglie's own abandonment of his pilot wave theory, physics students in the late 1940s were generally fed similar myth-historical narratives describing the quantum revolution using a "rhetoric of inevitability" that made quantum orthodoxy seem ironclad.[12] These narratives omitted the pilot wave and any mention of the lively interpretive debates of the late 1920s. In Lederman's myth-historical framing of scientific progress, because pilot wave theory had lost out to the Copenhagen interpretation, it should be filtered out as "noise" and "vanish into the debris of history's footnotes."[13]

When Bell first encountered quantum mechanics in a seminar at Queen's taught by the physicist R.H. Sloane, he did what many philosophically minded students do: he questioned it. Sloane was an experimentalist, grounding his quantum seminars in experimental and applied techniques.[14] Similarly, the textbook he used after 1949, Leonard Schiff's *Quantum Mechanics*, takes an applied approach to quantum mechanics. A brief opening chapter discusses "The Physical Basis of Quantum Mechanics"; the orthodox interpretation of quantum theory is presented not as an open epistemological question, but a closed ontological truth. In Section 3 of the first chapter, Schiff presents both Bohr's complementarity principle and Heisenberg's uncertainty principle as unavoidable, foundational pillars of quantum mechanics.[15] The orthodox interpretation of quantum theory is not questioned. It is presented matter-of-factly as an introduction to the mathematical formalism and applications that comprise most of the text.[16]

Sloane taught his quantum seminars as he was taught—as an exercise in rote learning of mathematical formulas and their application to approximate real-world problems. As a result, Bell and his classmates were thoroughly immersed in what Sharon Traweek has appropriately termed a "pilgrim's journey."[17] When Bell repeatedly questioned the conceptual basis for the measurement problem and the notion of an arbitrary delineation between observer and the system to be measured, Sloane dismissed his fiery student's inquiries. A tense back-and-forth reached a climax when Bell shouted accusations of intellectual dishonesty at his professor. The experience left young Bell feeling frustrated and intellectually unsatisfied.[18]

After his altercation with Sloane, Bell went on a quest for a deeper understanding of the prevailing interpretation of quantum theory (see Figure 2.3). He eventually came across Max Born's book *Natural Philosophy of Cause and Chance*.[19] Based on Born's seven Waynflete Lectures at Oxford University in 1948, the book carefully explored the philosophical implications of quantum theory (see Figure 2.4). Born claimed explicitly that quantum theory had revealed the nature of the atomic world as fundamentally indeterministic but not acausal. Intended for a general audience, his discussion was nuanced, striving to distinguish clearly between concepts like causality and determinism.[20]

Bell considered Born's to be one of the most influential texts he had read, but in his 1982 paper he complains that Born failed to mention de Broglie's deterministic pilot wave interpretation. Born thoroughly discussed causality, determinism, and chance in light of quantum theory, but never grappled with the epistemological

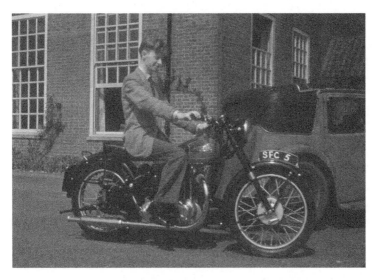

Figure 2.3 A young J.S. Bell getting ready to ride off on his motorcycle. (Source: Courtesy of CERN.)

Figure 2.4 Portrait of Max Born (Source: Courtesy of AIP Emilio Segrè Visual Archives, Gift of Jost Lemmerich.)

pluralism stemming from early quantum debates of the 1920s. Why did Born omit the pilot wave from his discussion of determinism? As this chapter will show, these were not the failures of a single author or the frustrations of a particular student.

Quantum Myth-Histories

The renowned physicist and philosopher of science James Cushing echoed Bell's frustrations with the omission of alternate quantum interpretations such as the pilot wave. In his book *Quantum Mechanics: Historical Contingency and the Copenhagen Hegemony*, Cushing regards this as a significant blind spot in the historical development of quantum mechanics. According to Cushing, "It is astounding that there is a formulation of quantum mechanics that has no measurement problem and no difficulty with a classical limit, yet is so little known. One might suspect that it is a *historical* problem to explain its marginal status."[21] More than simply a historical problem, the marginal status of this alternate interpretation of quantum theory is partly rooted in scientists' use of myth-historical narratives as filters.

The frustrations of Bell and Cushing hinge on the coordinated, intentional omission of a promising alternate interpretation of quantum theory and the general marginalization of interpretation debates by a powerful orthodoxy. Recent scholarship in the history of science has shown that the unity of the orthodox interpretation has been overstated. Scholars have declared the coherence of the orthodoxy to be a "myth."[22] The coherence of the Copenhagen interpretation may indeed be a myth, but the rhetorical power of the myth-historical narratives that emerged had real consequences for the physics community.

In order to understand the roots of the theoretical omissions and subsequent interpretational complacency that Bell and Cushing decried, we must first examine how the physics community reached consensus, canonized knowledge, and constructed its own myth-historical narratives beginning in the late 1920s.[23]

Historian Mara Beller has described how a small group of power-ful quantum physicists constructed their "revolutionary stories," fabricated "division between 'winners' and 'losers'," "misrepre-sented and delegitimized" their opposition, and sowed an illusion of consensus.[24] All this reinforced their "rhetoric of inevitability."[25] Recalling Bell's frustrations, it becomes clear that the rhetorical power of omission in orthodox quantum narratives had lasting consequences for generations of physicists and the science they practiced.

Although I agree with Richard Staley that "forgetting is integral to scientific advance," some scientist-storytellers intentionally employ myth-histories as rhetorical agents of omission in order to build consensus.[26] As in the case of Leon Lederman, scientist-storytellers are not necessarily engaged in a passive act of forget-ting.[27] The fate of pilot wave theory shows how canonical narratives can build social imaginaries, cement professional identity, establish scientific consensus, and influence current and future research.

To illustrate, we examine the early development of pilot wave theory and its later omission from quantum narratives. The story begins with de Broglie's development of a deterministic wave mechanical interpretation of quantum theory amid international schisms and interpretational flux after World War I. The Fifth Solvay Council in fall 1927 marked an important moment in the healing of international tensions and the building of scientific consensus around quantum theory. De Broglie presented his pilot wave theory at Solvay, but received very little encourage-ment from other participants. In particular, Wolfgang Pauli lev-eled sharp criticism. Toward the end of the council, some participants invoked a powerful rhetoric of closure to interpret-ational debates in an attempt to force consensus.

By 1930, de Broglie had abandoned pilot wave theory, and the rhetoric of closure was reinforced by rising pragmatism in the physics community. As the canons of quantum theory emerged, they omitted pilot wave theory and any reference to past inter-pretive debates. Over the next two decades, such debates retreated to the fringe of the physics community's discourse. Various reasons can be cited: the successful pragmatic applications of quantum mechanics; the demands of World War II; and rhetorical agents, such as von Neumann's impossibility proof against hidden variable theories. As a result, when Max Born wrote his popular quantum narrative in 1949, he omitted any mention of the interpretive debates of the 1920s, noting hidden variable theories only in passing.[28]

When Bell later learned more about quantum's nuanced history, he was frustrated by what he saw as clear rhetorical omissions of the pilot wave and interpretive debates. While I do not doubt that the rhetoric of myth-historical narratives played a critical role in building consensus and filtering out unwanted noise, we must be careful not to exaggerate their agency. As rational reconstructions of history, myth-historical narratives play complex roles in the dynamics of scientific communities. Whether you interpret them as catalysts to consensus or as the intentional agents of demarcation and omission, examining these narratives more closely may be fruitful, both historiographically and scientifically.

Emergence of de Broglie's Pilot Wave Interpretation

Although Arthur Eddington had warned his colleagues that using the pursuit of truth "as a barrier fortifying national feuds is a degradation of the fair name of science,"[29] World War I had a lasting effect on the international physics community. Due to post-war geopolitical tension, schisms arose within the scientific community that affected the reception of Louis de Broglie's pilot wave theory. Throughout the 1920s, important centers of quantum research and teaching grew at universities in Germany (Munich and Göttingen) and in neutral nations like Denmark (Copenhagen) and Holland (Leiden),[30] allowing a free exchange of ideas and personnel among them. But impediments to broader international scientific collaboration persisted, for German and Austrian scientists were barred from attending international conferences during the early and mid-1920s.[31] As a result, French scientists like de Broglie were limited in their exchange with German and Austrian scientists at these core quantum schools.

During this period, de Broglie (see Figure 2.5) began his doctoral research on electron orbits and the structure of atoms, working out of his older brother's X-ray spectroscopy laboratory.[32] De Broglie's scientific context was not only the front lines of international scientific conflict, but also the forefront of experimental explorations into wave-particle duality.[33] Though de Broglie was working on leading problems in quantum theory, he did so in relative isolation. Not only did international scientific schisms between France and the leading quantum schools make collaboration with the most influential quantum thinkers difficult, but the institutional structures and research traditions of French science also made de Broglie's research program unpopular even in France.[34]

Despite de Broglie's relative isolation from both the international quantum physics community and other physicists in France, two

Figure 2.5 Louis de Broglie sitting at his desk. (Source: Wikimedia Commons.)

influences decisively shaped his research. Experimental work being done in his brother Maurice's lab, as well as a series of papers by Marcel Brillouin on the hydrodynamic model of a vibrating atom, convinced de Broglie that a unified picture of radiation phenomena was possible. His goal was to reconcile Einstein's quantum hypothesis for radiation with Bohr's atomic theory.

Born's 1949 orthodox account of the quantum revolution describes de Broglie's contributions to the quantum revolution in flattering terms, yet his description seems a backhanded compliment.[35] Like most myth-historical accounts of quantum theory, Born's reduces de Broglie's contributions to the famous equation that bears his name and exemplifies wave-particle duality: $\lambda = h/p$.[36] These narratives tend to characterize de Broglie's doctoral thesis of 1924 as a flash of insight in which Einstein's quantum hypothesis is extended to the wave-particle duality of electrons. They fail to recognize the depth and potential of de Broglie's deterministic wave mechanical research program.[37]

De Broglie sought to develop a complete theoretical framework for electronic orbits in atoms by synthesizing both wave and particle pictures. His dissertation makes this point explicit: "We think that this idea of a deep relationship between the two great principles of Geometrical Optics and Dynamics could be a valuable guide in realizing the synthesis of waves and quanta."[38]

In 1969, V.V. Raman and Paul Forman published a paper titled "Why Was It Schrödinger Who Developed de Broglie's Ideas?" They argue that Schrödinger was the best positioned physicist in 1925–26 to develop de Broglie's ideas on the synthesis of wave-particle formulations.[39] The question in their title seems to imply that de Broglie failed to develop his own ideas. In fact, de Broglie did develop his own ideas. While his theory of wave mechanics differed from Schrödinger's, both shaped the landscape of intense speculation and innovation associated with quantum theory before 1927.

A closer examination of de Broglie's work shows that he continued to pursue a program of synthesis between the wave and particle pictures. Throughout the summer of 1926, while Schrödinger's more famous wave mechanical interpretation of the wave function (ψ) was being debated in Germany and Denmark, de Broglie was not idle. He began to publish a series of papers that developed his own wave equation approach to quantum mechanics.[40] This program would eventually evolve into his pilot wave deterministic interpretation of quantum theory.[41]

The difficulty in applying his method was that each particle would be affected by other particles' relative positions and trajectories. While it was sometimes possible to know the initial conditions of a system with one particle and therefore solve for its dynamical guiding equation, for general configurations of more than one particle, solving multiple, coupled partial differential equations quickly became mathematically intractable.[42] Reinterpreting the quantum wave function psi as a physically real "pilot wave" guiding the motion of microscopic particles was revolutionary; in theory, this interpretation could return determinism to quantum theory, but it contradicted much theory emerging from the main quantum schools in Germany and Denmark.

De Broglie's wave mechanical formulation of quantum theory was one of the more mature research programs, but he was not alone in trying to develop a single framework for periodic wave phenomena and classical deterministic particle trajectories. In the turbulence of the 1920s, physicists tried various approaches to reconcile quantum and classical physics.[43] From our perspective de Broglie's research program seems to stand out from the others for its coherence, persistence, and maturity.[44] Theoretical physics, today and then, is far from monolithic: it is a complex disciplinary fabric with many contributing threads. How and when each thread becomes celebrated or omitted is part of the field's nuanced history.

Fifth Solvay Council—Converging Lens

In the quantum narrative, the Fifth Solvay Council is traditionally assigned central importance as the battleground between scientific traditionalists and a new quantum ontology threatening to override physicists' classical interpretations of the natural world. While scientific debates surely occurred in Brussels during fall 1927, the conference itself became more a converging lens than a pitched battlefield. The council neither began nor ended a great philosophical battle. Instead it became a moment of rhetorical clarity for a scientific community embroiled in turbulent political and scientific divergence. As a leading statesman of the international physics community, Hendrik Lorentz was intent on political reconciliation, trying desperately to re-integrate German science into the international fold (see Figure 2.6). He was also intent on making the 1927 Solvay gathering a forum for scientific consensus on the relationship between classical and quantum theories (see Figure 2.7).[45]

Figure 2.6 Hendrik Lorentz in Leiden circa 1923 with Arthur Eddington (seated); standing, from the left: Albert Einstein, Paul Ehrenfest, and Willem de Sitter. (Source: Courtesy of AIP Emilio Segrè Visual Archives, Gift of Willem de Sitter.)

Figure 2.7 Official portrait of the 1927 Solvay Council (Brussels). Left to right, back row: A. Piccard; E. Henriot; P. Ehrenfest; E. Herzen; T. de Donder; E. Schrödinger; J.E. Verschaffelt; W. Pauli; W. Heisenberg; R.H. Fowler; L. Brillouin. Middle row: P. Debye; M. Knudsen; W.L. Bragg; H.A. Kramers; P. Dirac; A.H. Compton; L. de Broglie; M. Born; N. Bohr. Front row: I. Langmuir; M. Planck; M. Curie; H.A. Lorentz; A. Einstein; P. Langevin; C. Guye; C.T.R. Wilson; O.W. Richardson. Absent: Sir W.H. Bragg; H. Deslandres; E. Van Aubel. (Source: Photograph by Benjamin Couprie, Institut International de Physique Solvay, courtesy AIP Emilio Segrè Visual Archives.)

As in previous Solvay Councils, presentations began by surveying the current state of experimental techniques and results associated with the topic of discussion. Presentations on experimental techniques given at Solvay show that the foundations of quantum theory were thoroughly grounded in experimental observations. The rising power of the Schrödinger wave equation and Heisenberg's matrix mechanical formalisms were due to their success in application, not to an arbitrary decision to make them canonical.

Based on successes in applying quantum formalisms to explain experimental observations, some theorists—including Born, Bohr, and Heisenberg—felt emboldened to make somewhat premature, overly dogmatic statements of their interpretations of quantum theory. Subsequent Solvay presentations covered the theoretical foundations of quantum mechanics. De Broglie's paper summarizing his pilot wave picture of quantum mechanics came first.

Although critiques and discussions immediately followed de Broglie's paper, there were no clear disqualifying objections to its validity as a contending interpretation of quantum theory. Schrödinger had yet to give a paper with his own interpretation of quantum theory. So, as Born and Heisenberg began to give their paper thoroughly examining the core principles of matrix mechanics, the principle of indeterminacy, and the statistical interpretation of the wave function, there was no reason to think that the quantum interpretation debate had been settled.[46]

As a result, it seemed presumptuous to some when Born and Heisenberg wielded their rhetorical ax, trying to cut off quantum interpretational debates:

> By way of summary, we wish to emphasize that…we consider quantum mechanics to be a closed theory, whose fundamental physical and mathematical assumptions are no longer susceptible of any modification. [Furthermore,] the assumption of indeterminism in principle, here taken as fundamental, agrees with experience.[47]

This forced closure was clearly not the result of vigorous scientific debate at the Fifth Solvay Council. It was a rhetorical ploy based on arguments raging for quite some time. Amid the many political and scientific schisms of the 1920s, an alliance, on display at Solvay, had been forged between the two quantum schools in Copenhagen and Göttingen.

Forcing Closure of Interpretation Debates

As a retort to Heisenberg and Born's insistence that their presentation of quantum mechanics represented a closed theory, not susceptible to alteration with regard to its physical assumptions, Lorentz made explicit what seemed to be a growing point of tension among some physicists. During the general discussion section at Solvay, Lorentz commented that raising indeterminism to a fundamental principle seemed arbitrary and unjustified. The great Dutch scientist, in one of his last scientific acts, warned his colleagues that efforts to reconcile quantum theory with classical principles of determinism should not be abandoned.[48] Lorentz died three months later. It is clear that Einstein, Schrödinger, and de Broglie agreed with Lorentz's general critique of the closed vision of quantum theory that Born and Heisenberg presented, but the lack of a coordinated alternate position weakened their opposition.

Without clear, coordinated opposition, Bohr, Born, Heisenberg, and many of their colleagues left Solvay publicly insisting that quantum mechanics was now a closed theory, with indeterminism an unavoidable ontological principle of the new quantum reality (see Figure 2.8). The latest experimental evidence clearly

Figure 2.8 Werner Heisenberg and Niels Bohr in conversation over a meal in Copenhagen. (Source: Photograph by Paul Ehrenfest, Jr., courtesy AIP Emilio Segrè Visual Archives, Weisskopf Collection.)

showed distinct complementary pictures with no clear path to unification and reconciliation. In addition, the equivalent quantum mathematical formulations based on Schrödinger's wave mechanics and matrix mechanics were extremely effective in practice.[49] There seemed a growing sense that, in order to maximize the community's efforts to extend and apply quantum mechanics, the epistemological pluralism that had saturated the community in the years before the Fifth Solvay Council must congeal in consensus.[50] What emerged was a physics community that over the next two decades would be partly characterized by significantly more consensus and pragmatism, with a corresponding decline in quantum interpretive debates.[51]

A Rising Pragmatism, Consensus, and Interpretational Lull

As the 1927 Solvay conference approached, tensions arose between a growing group of pragmatists who wanted to attack and solve definite problems, and, as Oliver Lodge noted, "philosophically minded interpreters" who were introducing "[temporary] absurdities...into their scheme of reality."[52] Charles Galton Darwin, grandson of the celebrated biologist, exemplified the rising tide of pragmatism. In December 1926, Darwin wrote to Niels Bohr as follows:

> It is part of my doctrine that the details of a physicist's philosophy do not matter much.... Because the best sort of contribution that people like me can make to the subject is working out of problems, leaving the questions of principles to you. In fact even if the ideas on which the work was done are wrong from the beginning to end, it is hardly possible that the work itself is wrong in that it can easily be taken over by any revised fundamental ideas that you may make.[53]

Darwin's reticence to engage Bohr in any form of interpretive debate was not uncommon. On this point, he aligned with a growing silent majority of practicing physicists. Einstein, too, noticed more interpretational complacency in the physics community (see Figure 2.9). In a May 1928 letter to Schrödinger, he complained that the "Heisenberg-Bohr tranquilizing philosophy—or religion?—...provides a gentle pillow for the true believer from which he cannot very easily be aroused."[54]

If Darwin represented the pragmatic majority on issues of quantum interpretation, a vocal and powerful minority consisted, in part, of a group that John Heilbron called the "the earliest missionaries of the Copenhagen spirit."[55] The use of the term "missionaries" reflects Heilbron's conclusion that key quantum figures who helped shape the field early on could almost be considered religious zealots in how fervently they spread the principles, which eventually became synonymous with the Copenhagen interpretation. Heilbron even describes Bohr's most passionate supporters, like Pascual Jordan and Léon Rosenfeld, as "disciples."[56]

Heisenberg was not Bohr's most ardent "disciple" on notions of interpretation, but he was a leading apostle of this early Copenhagen spirit. He tentatively supported Bohr's complementarity principle

Figure 2.9 Niels Bohr and Albert Einstein deep in conversation. (Source: Photograph by Paul Ehrenfest, courtesy AIP Emilio Segrè Visual Archives.)

in fall 1927, but by summer 1928, he boldly declared that the "fundamental questions [in quantum theory] had been completely solved" and that Bohr's complementarity was the capstone of the whole interpretation.[57] As a result, while touring the U.S. during spring 1929, Heisenberg began talking of a "Kopenhagener Geist der Quantentheorie" or a "Copenhagen spirit of quantum theory."[58] The dissemination of this Copenhagen spirit widened when physicists like Heisenberg found themselves and their ideas reaching beyond continental Europe through lecture tours, voluntary and forced immigration, and publication of new quantum textbooks. After participating in the American Physical Society annual meetings held in Chicago in 1933, Bohr gave a speech at the World's Fair introducing his principle of complementarity to a popular audience.[59]

We must be careful not to characterize this "Copenhagen spirit" as a unitary interpretation or a singular front. Though Heisenberg and others generally subscribed to it, they certainly did not see eye-to-eye on all interpretive matters.[60] More accurately, the dissemination of the Copenhagen spirit via Heilbron's "missionaries" represented what Kenji Ito refers to as a "resonance" among the core set of quantum theorists.[61] Yet this Copenhagen spirit cannot be overlooked, for it became a powerful agent of demarcation within the physics community.

A careful reading of early correspondence among many architects of the quantum revolution led historian Mara Beller to conclude that the very notion of a singular quantum consensus known as the Copenhagen interpretation was illusory, a myth propagated for the sake of social control.[62] More recently, scholars such as Don Howard and Kristian Camilleri have deepened this analysis around the mythological origins of the Copenhagen consensus. As Howard notes, "The image of a unitary Copenhagen interpretation is a postwar myth, invented by Heisenberg [in 1955]. But once invented, the myth took hold as other authors put it to use in the furtherance of their own agendas."[63] Historical evidence shows that there was no such thing as a singular

Copenhagen interpretation of quantum theory in the 1930s and 1940s, yet one wonders if we are not minimizing an important historical actor. While Beller refers to the "illusion of the existence of a paradigmatic consensus,"[64] the illusion itself served as a powerful historical agent of demarcation long before Heisenberg's naming the "Copenhagen interpretation" in 1955.

Although it was not Heilbron's intent, an image of Copenhagen missionaries single-mindedly proselytizing a uniform quantum gospel according to Bohr has become, for some, the story of a hegemonic force of marginalization.[65] This characterization is not supported by historical evidence. Silvan Schweber's sweeping study of the development of quantum electrodynamics, *QED and the Men Who Made It*, reconstructs a community of physicists from the late 1920s to the late 1940s, debating each other as they struggled to apply, extend, and overcome inadequacies in the quantum formalisms established in the late 1920s. In particular, they struggled to extend the canonized quantum formalisms of Heisenberg, Schrödinger, and Dirac to account for the effects of special relativity.[66] In 1929, Dirac observed:

> The general theory of quantum mechanics is now almost complete.... The underlying physical laws necessary for the mathematical theory of a large part of physics and the whole of chemistry are thus completely known, and the difficulty is only that the exact application of these laws lead to equations much too complicated to be soluble.[67]

The formulations of quantum theory, he thought, represent an almost complete theory with only extensions, applications, and fixes left undone.

The narrative Schweber tells, however, shows dual modes of discourse within the quantum physics community during the 1930s and 1940s. Around topics focused on applying and extending the formulations, we see an open, vibrant debate prompting many attempts to complete the QED synthesis. Yet on issues of interpretation, there seemed little interest in publicly challenging what was codified as the Copenhagen spirit.[68] The second of these dual modes is what Bell faced in his Belfast training during the late 1940s. The process of interpretational codification thus occupies us here.

Canonizing Quantum Orthodoxy

Beginning with the first wave of quantum textbooks that appeared in the late 1920s and early 1930s, a concerted push began to encourage presentation of quantum mechanics as a closed theory that required a new philosophical approach to physical phenomena. These texts tended to emphasize the empirical foundations of the new theory and its practical power of application.[69] While they generally briefly discussed philosophical foundations of the new indeterministic ontology, they did so while omitting references to past interpretive debates. For example, Sommerfeld's 1929 edition of *Atombau und Spektrallinien* readily accepts Born's statistical interpretation of the wave function and Heisenberg's principle of indeterminacy and deals extensively with the measurement problem—but never mentions Bohr's principle of complementarity or any of its philosophical implications. Sommerfeld warns his readers that "I have essentially limited myself to such problems as can claim immediate physical interest," implying that complementarity had no practical application.[70]

As Suman Seth argues in *Crafting the Quantum,* the Sommerfeld school in Munich was very influential in constructing the practice of theoretical physics associated with modern quantum theorists such as Pauli and Heisenberg. Sommerfeld was an early pragmatist who moved away from concrete models to emphasize building empirical laws based on observables. He stood in direct contrast to the theoretical approaches associated with Planck, Einstein, and Bohr, who tended to pursue more general foundational principles. As a result, Sommerfeld's emphasis on empirical phenomena and his intentional avoidance of metaphysical lines of inquiry have a longer history, and, Seth argues, greater impact on the rising pragmatism of the 1930s than previously considered.[71]

At the time, focus on empirical and physically relevant problems was certainly justifiable, for a growing number of extraordinary successes occurred in applying various formulations of the new quantum theory. As Heisenberg pointed out, "The principles of the quantum theory have all been discussed, but a real understanding of them is obtainable only through their relation to the body of experimental facts which the theory must explain."[72] Indeed, a survey of forty-three textbooks on quantum physics published between 1928 and 1937 found that only eight mentioned the complementarity principle, whereas forty examined Heisenberg's uncertainty principle.[73] This is important. If Bohr's complementarity principle was not always an explicit part of the prevailing narrative propagated as the Copenhagen spirit, what chance did de Broglie's pilot wave have of being included in the quantum canon?

This was especially true because de Broglie, the pilot wave's most vocal and dedicated advocate, had abandoned his theory by 1930. The reasons for abandonment are complex and should not be reduced to extremes of individual agency or forced marginalization. To understand de Broglie's abandonment of the pilot wave, we can classify his reasons for conversion into three somewhat overlapping categories: personal, scientific, and social. The following analysis builds from the personal to the social, as I consider this last category to have most influenced his retreat.

De Broglie's Abandonment of Pilot Wave Theory

While de Broglie's personal context and temperament were not necessarily decisive in his abandonment of pilot wave theory, as part of a complex fabric of interwoven factors they are worth considering. The French physicist of noble birth had been working diligently and in relative isolation for years on his program to bring a unified, deterministic quantum theory to the table. Yet after de Broglie met resistance in 1927 on one of the largest scientific stages, he seemed to recoil. De Broglie's international star was rising, but he had not yet secured himself professionally in France. His dissertation work had been disseminated internationally by Paul Langevin, Einstein, and Schrödinger, and verified experimentally by Clinton Davisson, Lester Germer, and G.P. Thompson. He had been invited to present his latest research at the celebrated Fifth Solvay Council; soon he would be awarded the Nobel Prize. Yet not until 1928, with creation of the Institut Henri Poincaré (IHP), did de Broglie secure his first professional appointment as a lecturer in France.[74] Just as he was embarking on an academic career, de Broglie's timid, cautious nature seemed to overshadow his ambition.

De Broglie's work in isolation cannot be fully accounted for by international geopolitical tensions. His biographers have portrayed him as a passive, timid loner. His temperament was at least partially responsible for de Broglie not commanding a devoted following or quantum school, as did Sommerfeld and Bohr.[75] What some have interpreted as de Broglie's forced isolation and marginalization due to his theory might be read as self-inflicted isolation due to temperament and personality.

Another important factor in his abandonment of the pilot wave theory was that although de Broglie remained convinced that the program was viable in theory, he was well aware that his latest efforts were not a practical alternative to the other formalisms being adopted widely in the physics community. His approach, which required solving multiple, coupled differential equations, was feasible in principle, but solutions could be achieved only for contrived and simplified systems.[76] After seeing firsthand the powerful practical effectiveness of applying Schrödinger's wave mechanical and Heisenberg's matrix formalisms to real-world problems, de Broglie realized that pursuing his deterministic research program was no longer justifiable.

In an address to the meeting of the British Association in Glasgow in 1928 titled "Wave Mechanics and Its Interpretations," de Broglie unequivocally stated:

> The physical interpretation of the new Mechanics remains an extremely difficult subject. At the same time one important fact has been thoroughly established, viz. that both for Matter and for radiation we must assume the dualism of waves and corpuscles.[77]

His deterministic research program, seeking to unify wave and particle pictures of atomic phenomena, was nearly abandoned. Then, in his 1930 book, *An Introduction to the Study of Wave Mechanics*, he was explicit about orphaning his deterministic research program: "It is not possible to regard the theory of the pilot-wave as satisfactory."[78] By the time de Broglie compiled his book on *Matter and Light* for publication, he had added a note to his 1928 Glasgow address that emphasized indeterminism's near-universal adoption by the physics community: "Since this was written [in 1928], the interpretation of the New Mechanics in terms of Probability by Bohr and Heisenberg has been practically universally adopted by theoretical physicists."[79]

Finally, most importantly, we should again acknowledge the social role the larger physics community played in de Broglie's abandonment of his pilot wave theory. As noted earlier, the rhetoric of closure espoused after the 1927 Solvay Council became conflated with a resonance of consensus that coincided with a rising tide of pragmatism among physicists. In this intellectual environment, the once-fertile soil of quantum interpretation debates in the 1920s gave way to a discernible lull in interpretive debates in the 1930s and 1940s.[80]

From de Broglie's writings, we know he had personally embraced indeterminism as an inevitable reality and begun to recognize the larger community's acceptance of the basic principles of the Copenhagen spirit. Yet the historical evidence does not show that de Broglie was a victim of a hegemonic conspiratorial marginalization. It does show a vigorous effort by some of the community's core set of power brokers to ratify a particular understanding of quantum theory and unburden themselves of alternate interpretations that might divert attention from the growing successes of the theory's pragmatic applications. While we should not ignore de Broglie's individual agency in choosing to abandon his pilot wave theory, his decision can only be understood within the larger context of the quantum canonization in the international physics community in the early 1930s. After all, there were no other serious attempts to engage with the pilot wave theory between 1930 and the late 1940s, when J.S. Bell first encountered quantum mechanics at Queen's University Belfast.

Although the interpretive lull predated World War II, the effects of this global conflict reinforced the prevailing complacency of the international physics community toward interpretation. With the war effort demanding attention to refining immediate applications of their research, many physicists had little time to consider interpretations of quantum theory. The pragmatist mantra to "shut up and calculate" could be interpreted as an existential rallying cry during World War II. After the war, with the onset of the Cold War and the rise of the military-industrial complex, inertia and funding agencies kept many scientists focused on pragmatic applications of quantum theory.[81]

The Rhetorical Rise of Impossibility

Beyond the congealing consensus, growing scientific pragmatism, geopolitical tension, and loss of a personal advocate, another factor helped ensure the pilot wave would be omitted from quantum myth-histories. In the early 1930s, scientist-storytellers began using a rhetoric of impossibility to dismiss the pilot wave theory—and all deterministic theories like it—as viable alternatives to the prevailing quantum orthodoxy. This section examines the rhetorical rise of the von Neumann impossibility proof, which became an important calling card for myth-historians like Born, who wanted to justify omitting the pilot wave and rationalizing the inevitability of quantum orthodoxy.

Shortly after de Broglie abandoned his deterministic synthetic research program, John von Neumann published his seminal text on the mathematical foundations of quantum theory,

Mathematische Grundlagen der Quantenmechanik.[82] In many respects, this 1932 landmark text represents the apex of the axiomatization of quantum theory; prevailing formulations and interpretations were locked down into a canonized "complete" theory. Chapter Four of von Neumann's text argues for quantum theory's completeness by formalizing what became widely known as the "impossibility proof" against hidden variable theories:

> We need not go any further into the mechanism of the "hidden parameters," since we know that the established results of quantum mechanics can never be re-derived with their help.... The present system of quantum mechanics would have to be objectively false, in order that another description or the elementary processes than the statistical one may be possible.[83]

In other words, the indeterministic statistical formulations of quantum theory are unavoidable because it is a logically closed theory. It cannot be altered or reformulated in any way to re-establish determinism by introducing new, unobservable "hidden parameters." All deterministic alternatives, including the pilot wave theory, are therefore deemed impossible.

Although von Neumann's impossibility proof represented the apex of quantum axiomatization, as Trevor Pinch has noted, the heaviest impact of his proof was not in how it was used concretely within physicists' research programs, but how some authorities within the community invoked it in constructing their official histories.[84] It is telling that, in the years immediately after publication of von Neumann's book, the section on completeness seems to have been largely overlooked by most pragmatically oriented physicists. In this environment, many dismissed the axiomatic rigor in von Neumann's approach to quantum theory. Born went so far as to associate an axiomatic approach to physics with dogmatism, "the worst enemy of science":

> [Being] conscious of the infinite complexity he faces in every experiment [the physicist] refuses to consider any theory as final. Therefore—in the healthy feeling that dogmatism is the worst

enemy of science—he abhors the word "axiom" to which common use clings the sense of final truth.[85]

Although many in the physics community failed to directly grapple with von Neumann's text, some Copenhagen missionaries
did not hesitate to casually invoke the von Neumann impossibility proof if it served their purposes.[86]

Ironically, while years earlier Born had rejected axiomatic
approaches like von Neumann's, one sees the rhetorical power of
myth-historical filtering when reading Born's 1949 book on cause
and chance. He cites von Neumann's "brilliant book" in which
the great mathematical physicist

> puts the [quantum] theory on an axiomatic basis by deriving it from
> a few postulates of a very plausible and general character.... The
> result is that the formalism of quantum mechanics is uniquely
> determined by these axioms, in particular no concealed parameters
> can be introduced with the help of which the indeterministic
> description could be transformed into a deterministic one.[87]

Born's celebration of von Neumann's axiomatic impossibility
proof as a cornerstone of the quantum revolution seems a clear
case of winners writing myth-histories. Although Born had been
a respected, powerful architect of the quantum revolution since
the 1920s, he had not yet joined Bohr and Heisenberg in being
awarded the Nobel Prize. In 1954, his contribution of "the statistical
interpretation of quantum mechanics" was finally recognized by
the Nobel Committee for Physics. Knowing what we know about
the power of myth-historical narratives to shape community
dynamics, one wonders how much prevailing quantum myth-
histories, like Born's, that elevated quantum orthodoxy and filtered out alternate interpretations played in this recognition.

David Bohm Climbs Mt. Impossible

In the same 1982 paper in which Bell complained bitterly about
the missing pilot wave theory, he recounts a moment of exhilar-
ation. After a few years working as a practicing physicist, Bell

recalls seeing "the impossible done. It was in papers by David Bohm. Bohm showed explicitly how parameters could indeed be introduced, into nonrelativistic wave mechanics, with the help of which the indeterministic description could be transformed into a deterministic one."[88]

Of note here is Bell's playful use of Born's own rhetoric from 1949. By adding the word *indeed*, Bell inverts Born's meaning: "... how parameters could *indeed* be introduced, into nonrelativistic wave mechanics, with the help of which the indeterministic description could be transformed into a deterministic one."[89] It's clear that Bell found Born's erasure of de Broglie's pilot wave and his rhetorical use of von Neumann's impossibility proof in order to elevate "vagueness, subjectivity, and indeterminism" as hugely problematic and self-serving.[90]

In claiming he saw the "impossible done," Bell undoubtedly refers to David Bohm's now-famous work on hidden variables, first published in 1952 as a set of two papers.[91] In them, Bohm (see Figure 2.10) took aim at quantum orthodoxy by directly challenging

Figure 2.10 Portrait of David Bohm. (Source: Photograph by Mark Edwards. Courtesy Mark Edwards and Archive and Special Collections at Birkbeck, University London.)

von Neumann's impossibility proof and the accepted pillars of quantum interpretation. By modifying the Schrödinger wave equation, Bohm independently developed a deterministic formulation of quantum theory. Like de Broglie before him, he reinterpreted the wave function (ψ) as more than a mathematical entity with no physical meaning. Instead, Bohm conceived it as a "mathematical representation of an objectively real field."[92] Using this interpretation, and given initial positions and momenta, particle trajectories could be fully determined. Bohm had unknowingly reinvented the pilot wave theory.

Like Bell, Bohm had been blinded by the myth-historical erasure of the rich discourse around deterministic interpretations from the 1920s. He had no knowledge of de Broglie's groundbreaking work on pilot wave theory. Given this omission from myth-historical narratives, Bohm was shocked at the reception that followed his hidden variables publications in 1952. The context surrounding Bohm's climb up Mt. Impossible is critical to understanding this reception.

On December 3, 1950, the day before he was arrested by federal marshals on eight counts of contempt of Congress,[93] Bohm had a promising academic career path somewhat parallel to Richard Feynman's. In graduate school, he had made important contributions to the Manhattan Project and was considered by some to be "one of the ablest young theoretical physicists that [J.R.] Oppenheimer ha[d] turned out."[94] After the war, he settled in as an assistant professor of physics alongside Oppenheimer and other luminaries at Princeton University.[95]

In June 1947, the American Academy of Sciences sponsored the now-famous Shelter Island Conference. Modeled in part after the Solvay Councils, this elite conference was designed to give a limited number of the brightest young physicists, among them Bohm and Feynman, an opportunity to exchange ideas about the foundations of quantum theory in a relaxed, informal setting.[96] Before the conference, Victor Weisskopf, excited about the prospect of re-engaging his fellow physicists on the foundations of quantum theory, explicitly referred to the lull in interpretive activity that had gripped the quantum physics community for two decades:

> It is a good sign that somebody is again interested in discussing the foundations of quantum mechanics instead of thinking only of high voltage machines and how to produce mesons. The whole idea of a few quiet days in the country together with Heisenberg's "Uncertainty Relations" seems to me extremely attractive, and reminds me of wonderful days twenty years ago in Göttingen or Copenhagen.[97]

Bohm was one of the young physicists intent on addressing persistent issues with the foundations of quantum theory. Feynman and others were busy working out the divergence problems in quantum electrodynamics by using renormalization techniques. Their work was intended to fix and extend the accepted formulations of quantum mechanics, well within the framework of the Copenhagen spirit. Eventually, for this work, Feynman would share the 1965 Nobel Prize in physics with Sin-Itiro Tomonaga and Julian Schwinger.[98] On the other hand, Bohm's hidden variables program became, in many ways, a much more ambitious endeavor; he challenged von Neumann's impossibility proof and some core postulates of the orthodox interpretation. Instead of fixing and extending the applications of quantum mechanics, Bohm's research sought to reset the underlying foundations.

Bohm was a brilliant young academic, building a pioneering plasma research program and surrounded by dedicated students who found his teaching and mentorship invaluable. His quantum seminar had become very popular among Princeton physics students.[99] Preparing his lectures, he realized that existing textbooks were not ideal. Those like Schiff's, focused almost exclusively on applications, while Dirac and von Neumann's, which took a rigorous axiomatic approach to quantum, failed to grapple with the theory's physical implications. Bohm decided to write his own textbook.[100]

At the time of his arrest, he was completing final proofs for his textbook, *Quantum Theory*, published in the spring of 1951.[101] The book was well-received and adopted by many physics professors. Some of the most ardent supporters of the Copenhagen spirit found it one of the best presentations of their orthodox interpretation. But Bohm's textbook was far from conventional. It includes extensive discussions of the conceptual and philosophical foundations of quantum theory before even suggesting concrete applications.[102] It is true that Bohm seemed to side with orthodox interpretations and even argued against deterministic alternatives. However, a closer reading of the language and context shows ambivalence. His approach was open and skeptical. It contained seeds for what would grow into his hidden variables theory.[103]

Although Bohm's book extensively discusses underlying issues of interpretation and meaning, this was no guarantee that students of the early 1950s actually engaged in such discussions. A telling anecdote comes from an exchange between Einstein and two visiting graduate students in the early 1950s. After the students told Einstein they were using Bohm's textbook in their quantum seminar, he asked them how they "liked Bohm's treatment of the strangeness" implied by quantum theory. The students seemed stumped, and "couldn't answer." After all, they had "been told [by their professor] to skip that part of the book and concentrate on the section titled 'The Mathematical Formulation of the Theory'." One of them later recalled that the "issues that concerned [Einstein] were unfamiliar to us."[104] They had used Bohm's unconventional textbook infused with its excursion into foundational issues of quantum interpretation, but had skipped those sections and jumped right into quantum's pragmatic applications!

Despite publication of his successful textbook, 1951 was a tumultuous year for Bohm. After his arrest, Princeton's president placed Bohm on indefinite paid leave and forbade him to teach or visit campus until the trial was over. Forced into isolation, Bohm was given refuge, and a desk, at the nearby Institute for Advanced Study (IAS), where Oppenheimer was the director. Of course the most famous physicist at IAS was not Oppenheimer, but Einstein. During his brief stint at IAS, Bohm had long discussions with

Einstein about the foundations and interpretations of quantum theory. They discussed Bohm's textbook extensively and used it as a foil to challenge the orthodox interpretation. Soon, Bohm found inspiration to challenge the quantum orthodoxy he had just written about so convincingly in his textbook.[105]

Bohm's idyllic exile at IAS did not last. Although he was eventually acquitted of all charges of contempt of Congress, like many other victims of the House Un-American Activities Committee's (HUAC) persecution, he could never escape their repercussions on his fledgling career. He found himself unemployed and unemployable. Even Albert Einstein's glowing recommendations of Bohm were unable to overcome the stigma of the HUAC hearings.[106] By the time his orthodoxy-challenging hidden variables papers appeared in January 1952, Bohm had taken the only academic job he could find—in São Paulo, Brazil.[107]

The Rhetoric of Myth-Historical Responses

The interpretational lull and entrenchment in the orthodox understanding of quantum theory was such that in 1952 the physics community did not generally share Bell's appreciation of Bohm's work. Like Bell, Bohm had no idea that de Broglie and others had interrogated alternate deterministic formulations of quantum theory decades earlier; he was genuinely surprised when informed by de Broglie himself that he had tried this approach in the late 1920s but abandoned it. While Bohm thought he had broken ground with his deterministic interpretation of quantum theory, he was dismayed to find that his published papers met either bitter critique or apathetic silence. Although no one could disprove Bohm's proposed hidden variables interpretation, it met several biting rhetorical dismissals. De Broglie himself wrote to Bohm warning that a hidden variables approach, like the pilot wave theory before it, was a blind alley.[108]

One of the most biting responses to Bohm's papers came from Léon Rosenfeld, an ardent Copenhagen spirit missionary. In May

1952, Rosenfeld sent Bohm a patronizing letter using myth-historical rhetoric to categorically deny even the possibility that the orthodox interpretation could be reinterpreted in a deterministic way:

> I certainly shall not enter into any controversy with you or anybody else on the subject of complementarity, for the simple reason that there is not the slightest controversial point about it.... I may reassure you that...we have made probably all the errors that could conceivably be made before reaching at last solid ground. It is just because we have undergone this process of purification through error that we feel so sure of our results....I tell you this to show you that there is nothing dogmatic about our attitude of mind.[109]

Rosenfeld uses the myth-historical motif of "purification through error" to reassure Bohm that he, Bohr, and other missionaries are not in the least dogmatic about their entrenchment in the conventional understanding of quantum theory. Yet he opens the letter by dogmatically stating that there is not the slightest controversial point about complementarity. Rosenfeld's letter perfectly captures the rhetoric of inevitability and omission used by the historical winners of the quantum revolution to demarcate acceptable discourse and imprint their community's identity.

The reaction of Bohm's mentor, J.R. Oppenheimer, is also telling. A colleague visiting Princeton in January 1953 recalls hearing Oppenheimer bitterly refer to Bohm's hidden variables theory as "juvenile deviationism." He then added, "If we cannot disprove Bohm, then we must agree to ignore him."[110] As Born's Nobel Prize of 1954 and the coining of the "Copenhagen interpretation" the following year show, the physics community remained entrenched in the orthodox quantum interpretation. Many physicists dismissed any discussion of interpretation as metaphysics, unworthy of their time. For the most part, the community seemed happy to go about their business, shutting up and calculating.

However, a few theorists took note, and, like ripples in a pond, the effects eventually revealed themselves in surprising ways. As we have seen, Bell was very impressed by Bohm's bold takedown of the von Neumann impossibility proof. He quickly became an advocate for re-engaging quantum interpretation debates by developing empirical tests that could be used to falsify core postulates. Since the 1980s, testing Bell's inequalities has remained a central activity in research on quantum foundations.[111]

One of Bohm's earliest collaborators, Jean-Pierre Vigier, visited him in Brazil to work on extending his hidden variables theory. As it happens, Vigier was working as de Broglie's assistant in Paris, and eventually persuaded de Broglie to re-engage in the interpretation debate. While de Broglie had at first dismissed Bohm's 1952 papers, he eventually was rejuvenated as a determinist.[112] Like a slow cascade, other competing interpretations—including many-worlds, consistent histories, and objective collapse—eventually came into favor. By the 1990s, interpretive debates were again vigorous.[113] The pilot wave is no longer omitted. Its erasure has been reversed, reclaiming the theory from "the debris of history's footnotes."[114]

Conclusion

The history of the quantum physics community reveals an undeniable correspondence between how physicists see themselves and their history and how they go about teaching the next generation. This chapter has re-examined the context of the development and abandonment of de Broglie's pilot wave theory and explored the impact of its omission from myth-historical quantum narratives. The story of the pilot wave, and its erasure, shows that the consequences of propagating these narratives cannot be understood without confronting the forces that solidify consensus and canonize knowledge within a deliberately pragmatic scientific context.

Myth-historical narratives cannot be read solely as pedagogical interjections or entertaining anecdotes. They are powerful agents of demarcation that contribute to implanting standards by which the physics community functions. Explicitly or implicitly, they influence current and future research by delineating acceptable research questions from illegitimate lines of inquiry. While historians might aim to correct rhetorical distortions in these narratives, we must also understand that, among physicists, such distortions become effective pedagogical tools for carving out professional identity and scientific consensus, and thus play a powerful social role in science.

At the heart of John Bell's frustrations about omission of the pilot wave is a defiant challenge to both physicists and historians to acknowledge the power of propagating myth-histories. Considering Leon Lederman's discussion of myth-histories as rhetorical filters to remove unwanted noise, we should carefully note examples like the pilot wave theory. Although filtered and erased from pedagogical, scientific, and historical discourse, this alternate deterministic interpretation of quantum theory eventually returned to scientific conversation. Through its reincarnation as hidden variables theory, the de Broglie–Bohm theory, or Bohmian mechanics, the pilot wave theory has found a place among more than a dozen viable interpretations of quantum theory. Thoughtful juxtaposition of these might stimulate constructive cross-disciplinary dialogue between physicists and historians of science.

As John W.M. Bush discusses in "The New Wave of Pilot Wave Theory," researchers are currently exploring the fundamental characteristics of a walker-droplet system as a classical analogy to de Broglie's pilot wave quantum system. This finely curated macroscopic rotating hydrodynamical system seems to display dynamics similar to those proposed by de Broglie and his pilot waves in the late 1920s, including quantum effects such as nonlocality.[115] Some have renewed hope that we may eventually be able to do the impossible—bridge the chasm between microscopic and macroscopic domains, and reassess the place of indeterminism as an ontological truth. While the jury is still out on Bush's approach, we are reminded that sometimes forward-looking scientific vision works best by reflecting upon our past.

3

Myth-Historical CRISPR Edits

Modifying Futures by Controlling Pasts

There's nothing in the world more powerful than a good story.[1]

—TYRION LANNISTER

Heroes and Villains of CRISPR?

On January 14, 2016, one of the most influential scientists of the past quarter century published a short history of the CRISPR revolution in the prestigious journal *Cell*.[2] Read out of context, Eric Lander's narrative, "The Heroes of CRISPR," might be interpreted as an unbiased, well-researched history of a revolutionary gene-editing biotechnology. Yet his myth-historical narrative stirred up a controversial "shitstorm."[3] Responding to Lander's narrative, biologist Michael Eisen blogged that, by writing such dishonest, self-serving propaganda, Lander (see Figure 3.1) had morphed into the "villain of CRISPR." Eisen's comment made Lander's publication seem a plot twist from a James Bond film:

> There is something mesmerizing about an evil genius at the height of their craft, and Eric Lander is an evil genius at the height of his craft....I find it hard not to stand in awe even as I picture him cackling loudly in his Kendall Square lair, giant laser weapon behind him poised to destroy Berkeley if we don't hand over our [CRISPR] patents.[4]

According to Eisen, Lander's salvo reflects a pattern of distortion and dishonesty that has accompanied his rise in science, treating

Figure 3.1 Eric Lander giving a talk. (Source: Photograph by Maggie Bartlett; National Human Genome Research Institute, public domain. https://commons.wikimedia.org/w/index.php?curid=18157635.)

"truth as an obstacle that must be overcome on his way to global scientific domination."[5]

Does Lander's CRISPR history really rise to the level of trying to assert "global scientific domination"? How can a prominent scientist's attempt to write a history of a new biotechnology elicit such a melodramatic response? To make sense of Lander's historical narrative and the fallout it created, the following analysis interrogates his 2016 article as a myth-history. Because Lander's stated aim is to elevate certain heroes, his narrative falls squarely in the crosshairs of myth-historical analysis. As historians know, in history there are no absolute heroes or villains, only people with competing interests. When terms like "hero" and "villain" surface, as they do in Lander's essay and Eisen's response, it is a sure sign that myth-histories are being used to manage an underlying controversy. This chapter examines Lander's myth-history in comparison to a competing one written by Jennifer Doudna and Samuel Sternberg.

In Chapter 1 physicist Leon Lederman described these myth-histories as useful tools for scientist-storytellers to filter out of their narratives human foibles and controversies that they consider to be unimportant noise.[6] Ironically, the stories they tell about the CRISPR revolution are shaped by the very dynamics they omit. We are left with myth-histories that borrow heavily from idealized tropes of science, sketching simplified narratives in which presentist agendas dictate who is elevated to scientific heroism and who is not. Lander's narrative may seem a plausible reconstruction of CRISPR development, yet it glosses over a torrent of controversial factors complicating that story. Examining these filtered-out dynamics is essential to grasping how science really works. Properly framed, these myth-histories can reveal complementary perspectives and restore a more realistic view of the complex social and economic forces drawn to a promising new technology.

In another blogpost critiquing Lander, historian of science Nathanial Comfort is displeased that Lander wrote a "Whig history" of the development of CRISPR.[7] Although I agree with Comfort that Lander's piece is a presentist account of history that uses powerful rhetorical techniques to advance "a self-interested version of history,"[8] declaring it "Whiggish" allows historians to dismiss it as mere propaganda. Pitting scholarly history against Whig history leaves little room for fruitful discourse. Instead, treating Lander's piece as myth-history allows us to better unpack its context and compare it with competing myth-histories, such as that of Doudna and Sternberg. Each narrative is driven by a unique agenda with unexpected effects.

Lander is not a villain. He is a prominent scientist who happens to be an interested party in a contentious patent dispute. He interjected a myth-historical narrative, under the guise of history, to try to shape a preferred outcome to the ongoing controversy. However, by going beyond Eisen's dismissive response to examine the broader context of Lander's narrative, we can better understand its possible effects. Although the narrative works hard rhetorically to elevate some scientists and remove others from key roles, it also has corrective value. Lander's myth-historical narrative was not the first salvo, nor will it be the last, in the CRISPR controversy. Refusing the temptation to dismiss Whiggish accounts, we can move to a more reflective myth-historical epistemology.

Eisen's melodramatic blogpost sheds some light on Lander's context. As Eisen notes, part of Lander's motivation in writing his history seems to be a desire to influence the outcome of an ongoing patent dispute between scientists at Lander's home institution—the Broad Institute of MIT and Harvard—and Eisen's home institution—the University of California, Berkeley. Although Lander is hardly a Bond villain, his myth-historical salvo projects powerful rhetoric that could help decide who cashes in on CRISPR riches, including possibly billions of dollars in licensing fees, not to mention scientific immortality. Apart from personal egos and scientific truth, tremendous social, political, and institutional powers are at play in this conflict. We will examine them after summarizing the basic elements of CRISPR.

CRISPR Basics

Why is CRISPR technology so revolutionary that scientists struggle to write themselves into its history? The acronym CRISPR stands for clustered regularly interspaced short palindromic repeats. This refers to DNA sequences within prokaryotes, like bacteria, that mirror the DNA of invading viruses. CRISPR sequences can be understood as a prokaryote's evolutionary battle record, literally chronicling billions of years of viral attacks. Combined with CRISPR-associated enzymes that can cleave DNA and are precisely guided to their target by dedicated RNA molecules, the CRISPR system has evolved into a prokaryotic adaptive immune system that kills invading viruses. By studying exactly how the CRISPR defense system works, scientists have realized that they can transform it into a powerful, programmable gene-editing tool that can be inserted into a living organism and used in revolutionary ways (see Figures 3.2 and 3.3).[9]

Although other gene-editing nucleases like ZFNs and TALENs have been decades in development, and are being used in clinical therapeutic trials, they are expensive and difficult to manipulate. The power of the CRISPR revolution has been its capacity to disrupt and democratize gene-editing technology by making it widely accessible, relatively easy to customize, and comparatively inexpensive.[10]

CRISPR technology has already been used extensively to edit the DNA of plants and livestock,[11] and is now beginning to find uses in therapeutic contexts.[12] A recent list from a biotech news article reads like a menu from a science fiction story. Successful applications include breeding healthier pets, designing allergy-free foods, recording real-time events in the lifetime of a cell,

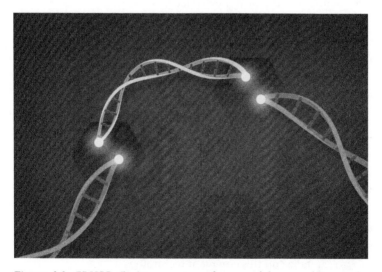

Figure 3.2 CRISPR-Cas9 is an extremely powerful gene-editing technology that stands to revolutionize biology with its precision and ease of use. (Source: Image by NIH Image Gallery from Bethesda, Maryland; CRISPR-Cas9, public domain. https://commons.wikimedia.org/w/index.php?curid=87951881.)

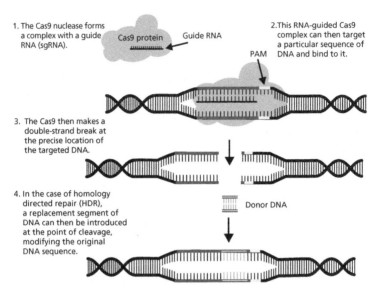

Figure 3.3 Schematic illustration of CRISPR-Cas9 gene-editing system. (Source: Illustration by Zeyu [Margaret] Liu.)

decaffeinating coffee beans, increasing biofuel efficiency, eradicating pests, and reviving extinct species.[13]

In addition, CRISPR has been featured in splashy headlines in both scientific and popular media. As one commentator noted, "CRISPR is the Model T of genetics."[14] High school students can buy CRISPR-based DIY kits for under $200.[15] The hype for this new biotechnology has grown so intense that, reportedly, in fall 2016 there were discussions about a new procedural TV drama titled "C.R.I.S.P.R." with Jennifer Lopez in the lead role.[16] With hopes, dreams, fears, and superlatives pinned on this revolutionary gene-editing technology, science fiction is quickly becoming science fact. Given all this, it is easy to understand a contentious race to capitalize on this emergent technology.

The first CRISPR-related patent application to target and edit DNA was filed with the U.S. Patent and Trademark Office (USPTO) in September 2008 on behalf of Erik Sontheimer and Luciano Marraffini at Northwestern University. That same year, Sontheimer's group had unambiguously shown that the CRISPR system tar-geted and cut specific nucleotide sequences of bacterial DNA. They had also recognized CRISPR's potential to be repurposed as a tool to cut and edit the genomes of other organisms. While the broad brushstrokes of this revolutionary technology had been painted, "a million mechanistic questions" did not allow these pioneers to fully describe, or reduce to practice, the CRISPR sys-tem. This technical threshold is a basic benchmark for the success of any patent application. Accordingly, Sontheimer and Marraffini's patent application was rejected by the USPTO in 2010.[17] Back then the CRISPR research community was still relatively small, limited mostly to microbiologists studying bacterial immune systems.

By 2012 the landscape had changed dramatically. The general roadmap to developing a revolutionary gene-editing technology existed, and many more labs around the world now engaged in CRISPR research. As research groups began independently inves-tigating the same terrain, the pace of discovery quickened, and CRISPR research became a battleground for priority disputes, patent controversies, and myth-historical interventions. In light of this early history of CRISPR research, we turn to examining Lander's 2016 myth-historical intervention in *Cell*.

Projecting a Miraculous Scientific Ecosystem

Lander opens his *Perspective* piece on "The Heroes of CRISPR" by arguing that scientists should take the history of their subject more seriously. He disapprovingly quotes Sir Peter Medawar's quip that most scientists find the history of science boring, and that, to many scientists, the circuitous details of scientific discov-ery are unnecessary distractions. Lander challenges this by claim-ing that "the human stories behind scientific advances can teach us a lot about the miraculous ecosystem that drives biomedical progress."[18] He believes that embracing this deeper, more realistic understanding of the scientific ecosystem is critical to ensuring that funding agencies, the general public, and students engage the scientific community in a constructive way.[19]

Lander claims that his motive for elucidating the history of CRISPR grew from seeing everyone debating the relative merits and potential of this advance in biomedicine without having any idea "*how* this revolution came to pass." In order to correct this, he claims to have spent "several months" studying the twenty-year backstory of the development of CRISPR. Lander relied on "published papers, personal interviews, and other written materials—including rejection letters from journals."[20] From this seemingly rigorous historical study of a particular scientific ecosystem emerges Lander's celebration of "a dozen or so scientists" and their collaborators, who together should be considered "heroes of CRISPR."[21]

Lander's claims resonate with many STS scholars and historians of science. Yes, scientists should be more aware of their actual histories and promote deeper understanding of the real work going on in their scientific ecosystems. Lander is careful to include many early pioneers of CRISPR research who might be brushed aside. He moves from the relative isolation of Francisco Mojica's work identifying CRISPR in the salt marshes near Alicante, Spain, to his own Broad Institute in Cambridge, Massachusetts, where, he claims, the CRISPR-Cas9 gene-editing system was finally operationalized as a revolutionary technology. Along the way, one infers that Lander has done his homework in historical research.

At first, a historian might even be encouraged by Lander's rhetoric of inclusivity. He positions himself as a disinterested champion of historical fact, glad to celebrate the uncelebrated for the sake of historical truth. Lander's narrative not only celebrates scientific accomplishments, it recognizes the struggles of the earliest CRISPR pioneers, like Mojica, to see their work published. Mojica's landmark paper identifying CRISPR as a bacterial adaptive immune system was at first rejected outright—without peer review—by the editors of journals such as *Nature* and the *Proceedings of the National Academy of Sciences*. After an "eighteen-month odyssey of frustration," Mojica's paper was finally published in the lower-tier *Journal of Molecular Evolution*.[22]

On closer inspection, Lander's account turns out to be a carefully crafted myth-history, infused with mythological tropes. Lander is not simply studying a run-of-the-mill ecosystem; the one he examines is a "miraculous ecosystem that drives biomedical progress." His essay is a curated presentist progress narrative that projects an image of science as an exceptional human activity. Lander clearly seeks to reconstruct the backstory of the CRISPR revolution and in the process to glorify certain "heroes." Since history itself is devoid of heroes, his task here approaches hagiographic construction. The tribulations experienced by scientists like Mojica create narrative tension that makes progress even sweeter.

Apart from brief historiographic platitudes promising to examine scientific ecosystems and sparse parenthetical comments about his heroes' frustration with barriers to publication and recognition, Lander seems to fall in line with the myth-historical ideals laid out by Lederman. CRISPR discoveries are linked in a clear causal chain of inevitability reminiscent of Lederman's myth-historical progress narrative of the physical sciences (see Figure 3.4). If Lander were really interested in unveiling a deeper understanding of the scientific ecosystem, he might have recognized the importance of Mojica's and others' publishing frustrations as a manifestation of deeper fault lines in a peer-review system that purports to be a neutral arbiter of scientific knowledge. Yet he never explores the broader impacts of systemic bias in scientific publishing and its peer-review systems.

That tangent might have led to interrogating more closely the social forces and topologies of power in science that have worked to the advantage of Lander and his peers. He might have been compelled to see the problems in power structures that favor certain privileged scientists while obstructing others at less "blue-ribbon" institutions. Had he taken this road, would Lander's scientific ecosystem still seem miraculous? Instead, he treats barriers, such as

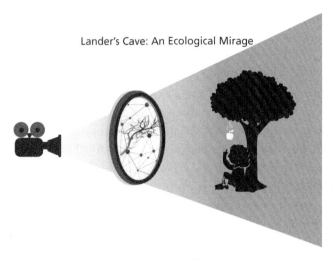

Lander's Cave: An Ecological Mirage

Figure 3.4 Lander's cave. Instead of Plato's cave, we see Lander's myth-history painting a "miraculous" scientific ecosystem. (Source: Illustration by Zeyu [Margaret] Liu.)

those faced by Mojica, as singular inconveniences—one-offs, not embeds in a flawed system. Blemishes are to be rhetorically exploited, then concealed or filtered out. No matter what their rhetorical goal, myth-historians like Lander and Lederman must ultimately preserve the image of a "miraculous" scientific ecosystem.

An Image May Be Worth Billions

Lander's myth-history recounts a step-by-step process of allegedly logical discovery. In order to build narrative tension and coherence, he divides the development of CRISPR into ten chapters that establish forward momentum. To illustrate his progressive causal chain, Lander inserts a map that frames CRISPR's creation as a continually unfolding narrative, propelled by scientific heroes; the story spans twenty years, twelve cities, and nine countries (see Figure 3.5).[23] The map depicts a powerful rhetorical timeline that plants the reader in a particular myth-historical framing of the technology's history.

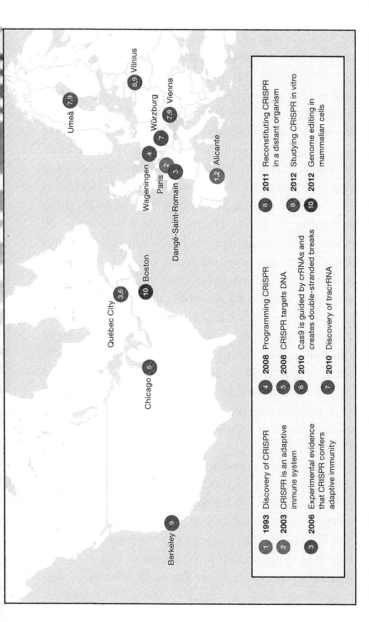

Figure 3.5 This map illustrating the development of CRISPR–Cas9 plays a central role in Lander's "Heroes of CRISPR" myth-history. In using this particular map representing the timeline for discovery, Lander makes some clear rhetorical choices that serve to frame the history in a particularly self-serving way. (Source: Reprinted from *Cell* 164, no. 1–2, Eric S. Lander, "The Heroes of CRISPR," 18–28, 2016, with permission from Elsevier.)

As Lander moves through these ten chapters—each identified by a color-coded, numbered circle—his narrative begins to seem inevitable. Lander's map offers clear signposts: the map's color codes are not merely design values; they perform rhetorical and narrative work. The first two chapters are green, the next seven are red, and the final one is blue. This signals a strategic periodization that emphasizes both the earliest work on CRISPR and the 2012 achievements of his colleagues associated with the Broad Institute. The emphasis is on the early and late periods; the intermediate steps, coded red, are minimized. Also, the word count in his narrative dedicated to each period reflects this emphasis directly. Of the ten chapters in Lander's narrative, three are either green or blue. Yet these three chapters account for almost half the publication. The remaining seven intermediate, red chapters are thereby squeezed and de-emphasized.

Within both green and red circles, some accomplishments are attributed simultaneously to multiple groups in Europe and North America. This reflects the geographical diffusion of scientific practice and the simultaneous, often redundant nature of research and innovation. Though only one green circle marks the first chapter, placed in Spain, two greens identify the second chapter, in Alicante and Paris. Moving to the red circles, this effect grows more pronounced: the seventh chapter occurs in three locations simultaneously, while the ninth chapter covers four distinct locations.

Thus, the configuration and signposts of Lander's map diffuse credit for developments in red-coded intermediate chapters. It is less than surprising that, in the final chapter, Lander elevates work associated with the blue circle as the culmination of all previous CRISPR research. The single blue circle is pinned on the greater Boston area, home of his Broad Institute.

Finally, to keep the spotlight on the work of his Broad colleague Feng Zhang, while downplaying the work of scientific competitors like Jennifer Doudna at UC Berkeley, Lander visually distorts and spatially recenters CRISPR's history. Readers can't avoid the blue circle directly in the center of the image. Why would someone place Boston at the center of a global map? Is that merely an artifact of using Google Maps from a location near Boston? Or is there rhetorical purpose here? As colonial powers have shown for millennia, a map is extremely effective in establishing, recording, and claiming power relationships between geographical locations and institutional affiliations.

Lander's map lays claim to power quite effectively, for its dimensions clearly distort space; the U.S. is much larger than it should be at that scale. On this map, Berkeley is almost two times farther from Boston than Paris is. Had the map been drawn to scale, the distance from Boston to Paris would be 25 percent longer than the distance from Boston to Berkeley. In effect, Lander's spatial distortion amplifies the isolation of Doudna's group by placing this lone red circular island against a sea of white. Berkeley's red circle is far indeed from the rest of the scientific action in Boston and Europe during this critical time for CRISPR. One last point about Lander's myth-historical map: where is the Global South? As we shall see, the international patent disputes raging over CRISPR include Jin-Soo Kim team's claims out of South Korea. Where are they?

Curiously, the tenth and final chapter describes work done in 2012, but Lander wrote this history in 2015–16. Did nothing really happen after 2012? Why are no milestones flagged after the work of his colleagues at the Broad Institute? Lander asserts that the 2012 work at Broad took "the final step of biological engineering to enable genome editing."[24] In a short aside after his chapter highlighting Zhang's revolutionary contributions to the CRISPR field, Lander claims that "Google searches for 'CRISPR' began to skyrocket" as interest in the revolutionary technology went viral.[25] Is this Lander's way of claiming foundational status in the patent dispute? Does he imply that all innovation after 2012 should be credited to Zhang?

On October 7, 2020, the Royal Swedish Academy of Sciences pushed back on Lander's myth-historical characterization of the development of CRISPR. Without explicitly mentioning Lander or his 2016 article in *Cell*, they decided to award the 2020 Nobel Prize in Chemistry solely to Doudna and Emmanuelle Charpentier for their "epoch-making experiment." In the press release, the Academy pointed out that since "Charpentier and Doudna discovered the CRISPR/Cas9 genetic scissors in 2012 their use has exploded."[26] Had the Nobel Committee created a map and timeline describing the development of CRISPR, Doudna and Charpentier's work probably would have received the blue circle. Although not an explicit rebuke of Lander's myth-history, awarding Charpentier and Doudna science's highest prize and calling their work "epoch-making" is telling.

Contested Patent Context

Lander's epic journey of scientific discovery omits a parallel untold story that continually informs his myth-history. It is a tale of power grabs, influence, scientific prestige, and economic riches. These tropes are not usually associated with the ideals of scientific myth-histories; no wonder Lander's tale omits them. Yet, when one grasps how close Lander is to the various priority claims and patent disputes that infuse CRISPR history, it is difficult to take his stated historiographic intent at face value. As a cofounder and director of the Broad Institute, his interests seem rooted more in supporting his colleagues and glorifying their accomplishments than in realistic description of the scientific "ecosystem." Although Lander may not be directly financially invested in the outcome of the patent disputes, as cofounder and director of the institute named as a party in the legal proceedings, he is certainly deeply entangled in the ongoing controversy.

Amid a caustic patent dispute, Lander's narrative and mapping are, at minimum, problematic. Regardless of intent, in January 2016, he was an interested party in the legal controversy and should have clearly acknowledged that. As one observer noted at the time, "The problem is, the Broad is a co-patentee embroiled in an intellectual property battle being investigated by the

U.S. Patent and Trademark Office (USPTO). And Lander's *Cell* paper does not disclose the potential conflict of interest."[27] Seen in this broader context, Lander's myth-historical narrative omits critical details such as the ongoing patent disputes involving his home institution.[28]

In 2012, biochemists had begun to isolate the essential components and underlying mechanisms of the CRISPR-Cas9 system. That spring, two independent research groups conducted critical in vitro experiments detailing workings of the prokaryotic immune system and unlocking the potential of this revolutionary gene-editing technology. One group was a joint collaboration between Jennifer Doudna's lab at UC Berkeley and Emmanuelle Charpentier's lab at the University of Vienna (see Figures 3.6 and 3.7). The second group was led by biochemist Virginijus Šikšnys (see Figure 3.8) from his lab at Vilnius University in Lithuania. Both groups identified what they considered the core molecular elements of the CRISPR-Cas9 system, and proposed

Figure 3.6 Jennifer Doudna. (Source: Photograph by Duncan Hull; own work, CC BY-SA 4.0. https://commons.wikimedia.org/w/index.php?curid=54379852.)

Figure 3.7 Emmanuelle Charpentier. (Source: Photograph by Bianca Fioretti, Hallbauer & Fioretti, copyright owned by Emmanuelle Charpentier who made it a Creative Commons picture, CC BY-SA 4.0. https://commons.wikimedia.org/w/index.php?curid=44041020.)

Figure 3.8 Virginijus Šikšnys. (Source: Photograph by Norwegian University of Science and Technology; KavliPrize-7021, CC BY-SA 2.0. https://commons.wikimedia.org/w/index.php?curid=72705093.)

harnessing it as an innovative, powerful, programmable gene-editing tool.[29]

As in most scientific innovations of this magnitude, attribution of credit for the discovery and invention has not gone smoothly. While few doubt that Doudna and Charpentier deserve significant credit for their trailblazing publication of June 2012, a pitched legal battle continues to this day over who controls the larger landscape of intellectual property associated with CRISPR.[30] At stake are claims to patents that may be worth billions of dollars in licensing fees as well as control of the development and commercialization of CRISPR-Cas9 technology.[31] Even with such high stakes, the ferocity and vitriol of CRISPR patent disputes have shocked many observers. As one headline from *Nature News* claimed, the "Titanic clash over CRISPR patents [has turned] ugly."[32] Legal proceedings have morphed from technical disagreements over particular scientific accomplishments to aggressive public accusations of impropriety.[33]

The purpose of any patent application is to temporarily protect an inventor against someone arbitrarily copying and using their invention. A patent can be an extremely powerful way to control and profit from intellectual property, so patent agencies like the USPTO must be judicious in reviewing the novelty and usefulness of any application. When the patent application describes an invention that builds or extends preexisting subject matter, it needs to be different enough to consider the invention "non-obvious to a person having ordinary skill in the area of technology related to the invention."[34] One key point of contention in the dispute over CRISPR patents seems to be the obviousness of extending the CRISPR-Cas9 gene-editing system from working on prokaryotic cells to applications in eukaryotic cells (found in organisms, like mammals, whose cells have a membrane-enclosed nucleus).

A flurry of activity in the CRISPR research community occurred in fall 2012, as multiple groups worked independently to extend the newly articulated CRISPR-Cas9 gene-editing system to work in vivo on other organisms, including eukaryotes. In January 2013, six major papers appeared that described these accomplishments. The first two papers, published simultaneously in *Science* on January 3, came from groups led by Feng Zhang (see Figure 3.9) at the Broad Institute and George Church at Harvard. Both described successful instances of in vivo homologous-directed repair of mammalian DNA using a modification of the CRISPR-Cas9 system described earlier in 2012 by Doudna, Charpentier, and Šikšnys. This work in Cambridge, Massachusetts, was the blue circle Lander was trying to claim as foundational in his CRISPR timeline. Another four papers appearing later in January 2013 verified and reiterated the extension of CRISPR-Cas9 to other cell types, including mammalian and human cells. The second wave of papers, appearing just a few

Figure 3.9 Feng Zhang with NSF Director France Córdova, circa 2014. (Source: Photograph by National Science Foundation; Annual Awards Recognize Outstanding Contributions in Research and Public Service, Public Domain. https://commons.wikimedia.org/w/index.php?curid= 65647608.)

weeks after Zhang and Church's, emerged from Doudna's group at Berkeley, Luciano Marraffini's at Rockefeller University, Keith Joung's at Harvard Medical School, and Jin-Soo Kim's at the Seoul National University in South Korea.[35]

Matching the breakneck pace of CRISPR innovation in 2012 and 2013, a barrage of patent applications tried to lay claim to this novel technology's potential. Among the first to file an application were Doudna, Charpentier, and their respective institutions.[36] On May 25, 2012, a patent application was filed with the USPTO on their behalf[37]—exactly two weeks before they submitted their landmark paper to *Science*; the application included "155 claims, encompassing numerous applications of the system for a variety of cell types."[38] The scope of their work and foundational patent application focused on a broad proof of concept in which Doudna and Charpentier reduced the workings of the CRISPR-Cas9 system and showed how it could be leveraged to precisely cut and edit prokaryotic DNA.

Then, on December 12, 2012, weeks before his paper was to appear in *Science*, Feng Zhang and the Broad Institute filed a patent application of their own.[39] Although Doudna and Charpentier's patent application had been filed almost seven months earlier, Zhang's application was submitted through an expedited USPTO review program. "Fast-track" patent applications require an added filing fee and are generally intended for undisputed applications. From Zhang's perspective, his team's expedited patent application didn't conflict directly with Doudna and Charpentier's because his invention appeared to take the nonobvious step of transforming the CRISPR-Cas9 system to use it in vivo on eukaryotic cells. In April 2014, the USPTO agreed with this assessment, granting the first patent for CRISPR-Cas9 applications to Zhang and his team.

Patent Interference

In the meantime, Doudna and Charpentier's original foundational patent application languished in USPTO bureaucratic limbo. At last, in 2015 they asked the USPTO to investigate the possibility that Zhang team's expedited patent claim had resulted in "patent interference." A hearing before the U.S. Patent Trial and Appeal Board (USPTAB) would decide which team first developed the technology, who had priority, and who should ultimately be awarded the patent.

On February 15, 2017, the USPTAB ruled in favor of the Zhang team's claim that there was no interference due to the fact that the two patent applications dealt with "patentably distinct subject matter."[40] After reviewing all the evidence, the USPTAB decided that extending the CRISPR-Cas9 system to eukaryotic cells was a nonobvious innovation and thus independently patentable. The Doudna-Charpentier team thought the USPTAB had relied on a problematic understanding of what defines a "nonobvious extension," and appealed the ruling to the U.S. Court of Appeals for the Federal Circuit in Washington, DC.[41] Their argument hinged on the fact that multiple labs independently, simultaneously, and quickly were able to extend their CRISPR-Cas9 gene-editing system by modifying the sgRNA and adapting it to work in eukaryotic cells.

The appeal of the USPTAB's decision was heard in late April 2018, and on September 10, 2018, the U.S. Court of Appeals affirmed the USPTAB's ruling. Based on a standard of "substantial evidence," the Appellate Court found that it was reasonable for the USPTO to award Zhang's team a patent in 2014. The scientific basis of this decision remains in dispute, yet, legally speaking, the outcome seems reasonable. As one legal scholar put it, there is a fundamental "disconnect between the legal standards of patent law and the realities of scientific research."[42] At the heart of the matter is a somewhat subjective measure of what constitutes an "obvious extension" of previous scientific work and what defines a patentable innovation.[43] The Royal Swedish Academy of Sciences decision to award the 2020 Nobel Prize in Chemistry to Charpentier and Doudna notwithstanding, there still seems no clear scientific consensus on how to attribute credit in this dispute.[44]

Although the particulars of the CRISPR-Cas9 patent interference dispute seem to have been resolved in court, more twists awaited the battling sides. On February 8, 2019, the USPTO issued a "notice of allowance" to UC Berkeley; they would grant the Doudna-Charpentier team a foundational patent based on their previous filings.[45] The imminent award of this patent, based on application no. 13/842,859,[46] is unrelated to the original interference proceedings, yet it will certainly spark renewed debate over who controls the future of the technology. This latest turn injects more uncertainty into an already fractured, contentious CRISPR-Cas9 patent and licensing process.

Complex Patent Landscape

Regardless of the long patent dispute, commercialization of CRISPR continues at a torrid pace. Key stakeholders from around the world are protecting their turf by applying to various patent jurisdictions like the European Patent Office (EPO) and the Chinese Intellectual Property Office (SIPO) to safeguard their CRISPR inventions on a global scale. As a result, the global CRISPR patent landscape is crowded and entangled. By one accounting, over four thousand patent applications have been filed worldwide on various applications involving CRISPR.[47] Of those, more than 2,300 patents have been awarded to thirteen distinct entities.[48]

Though Zhang's team has been granted dozens of U.S. and European CRISPR patents, many of their earliest European patents are being overturned on technicalities.[49] If the latest EPO decisions hold up, we may face a situation where the Doudna-Charpentier team holds broad-ranging foundational CRISPR patent rights in Europe, while the Zhang team holds identical rights in the U.S. As global CRISPR patent wrangling continues, it's difficult to predict how the convoluted landscape will settle. Based on the sheer number of overlapping patents worldwide, for the foreseeable future the global landscape for CRISPR will continue to be rugged.[50]

Such a patent storm is not unusual in industries that tie research and development directly to the potential commercialization of new technologies. What seems different about CRISPR is that the primary stakeholders are scientists, their academic institutions, and the well-funded start-up companies they have founded in order to control CRISPR development and reap the rewards of an impending gold rush. Since the Bayh-Dole Act of 1980, university-appointed scientists and their institutions have had a clear path to capitalize on patents stemming from their federally funded research. Before Bayh-Dole, rights to inventions arising from federally funded research were ceded to the government. Yet the controversy over CRISPR patents seems more divisive and vitriolic than with previous inventions from the past four decades of academic research. For example, with the Cohen-Boyer patents on recombinant DNA (rDNA) issued in the 1980s, Stanford University managed to avoid an ugly patent dispute by creating a system of licensing that has come to exemplify best practices for managing academic patents.[51]

Nearly four decades after Stanford negotiated a settlement among various stakeholders in the rDNA case, the CRISPR case has failed to meet this gold standard of shared inclusivity. Early on, stakeholders like Doudna and Charpentier claim to have favored a collaborative approach in which they tried to negotiate shared governance of CRISPR-Cas9 patents. However, for reasons not openly discussed, these early negotiations toward unified management of the patents "collapsed—with a good deal of noise and dust."[52] Entrenched business interests, different management styles, geography, and clashing egos may all have played a part in the collapse. In any case, since 2013, CRISPR patent battle lines have been drawn and rhetorical vitriol has flowed freely and publicly. In light of all this, let us return to Lander's myth-history.

Lander's essay was published in January 2016, just when the USPTAB was reviewing the patent interference claim lodged by UC Berkeley in 2015. With his proximity and timing, Lander seems to have aimed at bolstering one side of an active patent dispute.

Without knowing Lander's actual intent for this myth-history, we cannot label him an "evil genius," but in 2016 he was certainly an interested party who had enormous institutional wealth and prestige at stake in the controversy. In writing his myth-history for *Cell*, Lander seemed to clearly illustrate the idea that winners write history. The NIH, for one, is listening to Lander. In an overview of gene-editing and CRISPR-Cas9, they cite Lander's "Heroes of CRISPR" as an important "scientific journal article for further reading."[53]

Although Lander's map of CRISPR heroes is a distorted presentation of the development of this technology, we must also note the restorative effects of his framing. By periodizing and recentering the CRISPR map, Lander places the red circle representing the ninth chapter in four distinct locations, diffusing the importance of any one location and calling into question existing claims of priority and innovation. In so doing, his myth-history highlighted the work of Virginijus Šikšnys in Lithuania and became a catalyst for a publishing priority dispute.

Diluting Doudna and Charpentier

Despite Jennifer Doudna's considerable reputation within the field, Lander chose to significantly downplay the scientific accomplishments and contributions of both Doudna and Emmanuelle Charpentier. He briefly introduces Doudna in the ninth chapter, titled "Studying CRISPR *in vitro*."[54] This is the only red circle associated with four different research groups simultaneously. Curiously, of the heroes Lander mentions in this chapter, Doudna receives by far the least emphasis. She first appears in a paragraph midway through the section. While Lander does call Doudna a "world-renowned structural biologist and RNA expert," his description of her work in collaboration with Charpentier is almost completely eclipsed by his focus on the work of Šikšnys.[55] Šikšnys actually enters Lander's story in the previous section, "Reconstituting CRISPR in a Distant Organism." As Lander traces

Šikšnys's work on CRISPR from 2007 on, he ends this chapter by proclaiming the following:

> The field had reached a critical milestone: the necessary and sufficient components of the CRISPR-Cas9 interference system—the Cas9 nuclease, crRNA, and tracrRNA—were now known. The system had been completely dissected based on elegant bioinformatics, genetics, and molecular biology. It was now time to turn to precise biochemical experiments to try to confirm and extend the results in a test tube.[56]

As Lander transitions from Chapters 8 to 9, Doudna and Charpentier's in vitro work is treated as mere confirmation and extension of previous CRISPR research, not as the breakthrough and epoch-making discovery that many in the community have described. Lander's syntax implies that their accomplishments derive from Šikšnys: "Like Šikšnys, they showed that…"[57] Lander's narrative portrays all of Doudna and Charpentier's work as redundant to Šikšnys's work, thereby diluting their contributions.

If Lander's narrative has so far downplayed Doudna and Charpentier, the next paragraph further clouds their claims of priority and originality—claims critical to any patent. According to Lander, Šikšnys had written up his results and originally submitted his manuscript to *Cell* on April 6, 2012. However, within six days, "The journal rejected the paper without external review." After editing, Šikšnys resubmitted his paper "on May 21 to the *Proceedings of the National Academy of Sciences*, which published it online on September 4."[58] Lander's main point is not to question how the scientific community's peer-review system works or how priority claims are settled. As in his description of Mojica's publishing frustrations, there is no follow-up discussion. Lander is far more interested in noting pointedly that Doudna and Charpentier's landmark paper was "submitted to *Science* 2 months after Šikšnys'" and that, in contrast to Šikšnys's experience, their paper "sailed through review."[59]

As someone supposedly thinking deeply about the ecosystem of science and the stories behind discoveries, Lander deserves this point to be scrutinized. Unfortunately, he fails to pursue the topic. Without such discussion, it seems clear that Lander's primary aim for his myth-history was not to champion Šikšnys or outline the true social ecology of science. He sweeps all potential scientific controversy under the rug by generally minimizing the role of in vitro experimentation on CRISPR development. The in vitro work that showed the promise of using CRISPR-Cas9 to edit a genome had been accomplished on prokaryotic microbes; the big breakthrough would come by making this system work in mammalian eukaryotic cells:

> Since Marraffini and Sontheimer's 2008 paper showing that CRISPR was a programmable restriction enzyme, researchers had grasped that CRISPR might provide a powerful tool for cutting, and thereby editing, specific genomic loci—if it could be made to work in mammalian cells. But this was a big "if"....As late as September 2012, experts were skeptical.[60]

The "if" is critical here. In the patent interference dispute discussed earlier, the obviousness or nonobviousness of extending CRISPR-Cas9 from prokaryotes to eukaryotes was a key point. Lander's use of "if," followed by his description of experts as skeptical, underscores his support for the nonobviousness of this extension. In addition, his dating of expert skepticism to "as late as September 2012" is important. The landmark papers of both the Doudna-Charpentier and Šikšnys teams had been published by then. So, regardless of all the work by biochemists highlighted in the previous chapter, *nonobvious* innovations were still needed to transform CRISPR-Cas9 into a eukaryotic gene-editing tool.

The discussion here sets up Lander's most significant hero of the CRISPR revolution. He has highlighted what he considers a critical, nontrivial threshold in developing gene-editing technology. Lander dramatically points to the transition from using CRISPR-Cas9 in vitro with microbes to its far more revolutionary application to in vivo experiments on mammals. Lander now throws the spotlight directly on the Broad Institute's Feng Zhang. As he did with Šikšnys, Lander traces Zhang's scientific journey from childhood and makes sure to date his work on CRISPR well before the flurry of activity in 2012.

According to Lander, when Doudna and Charpentier's paper was published in late June 2012, it prompted a race to make CRISPR-Cas9 work on mammalian cells. Zhang immediately tested their system and found that their fused sgRNA worked poorly in vivo, "cutting only a minority of loci with low efficiency." Zhang realized that this problem could be solved by using "a full-length fusion that restored a critical 3' hairpin" code to the tracrRNA that dramatically reduced "off-target cutting." But Lander fails to recognize the other groups working independently and simultaneously on adapting the same sgRNA to work in eukaryotic cells. As noted earlier, apart from Zhang's efforts published in January 2013, five other publications claimed similar results that same month.[61]

In addition to making the CRISPR-Cas9 system operationally more effective and suitable for editing mammalian cells, Lander credits Zhang with broadening the scope and potential applications of the technology. However, most important to his rhetorical argument, Lander's narrative highlights Zhang's interest and research on the CRISPR-Cas9 system before June 2012. Thus Lander can confidently claim that Zhang's research on applications of the revolutionary technology to human cells paralleled the work of Doudna and Charpentier on microbes, and did not depend on their findings. Zhang's accomplishments from fall 2012 do not owe a great debt to his rivals' contributions. They can be understood as nonobvious extensions of Doudna and Charpentier's work, making them independently patentable. With this in mind, one should probably read Lander's narrative as partisan testimony in the ongoing patent dispute.

CRISPR Priority Disputes—Messy and Uneven Publishing Landscapes

The outcomes of Lander's myth-historical distortions are not all bad. There are also seeds of corrective justice, as he points to the messiness of uneven publishing landscapes. In pursuing scientific knowledge, tension has always existed between cooperation and competition. Although myth-historical ideals emphasize sharing scientific knowledge for the benefit of all, the system of recognition and rewards in scientific communities often bestows the honor on those considered first to achieve a goal. Whenever multiple research groups simultaneously and independently work toward a similar goal, a race ensues. As Lander noted in his myth-history, in the case of elucidating the CRISPR-Cas9 complex, Šikšnys's group was the first to write up and submit their results for publication. Yet Doudna and Charpentier were able to access an expedited pipeline to publication and ensure that their paper appeared first.

On April 6, 2012, Šikšnys submitted his paper to *Cell* for publication, but the editors quickly rejected it without distributing the paper for peer review. For the next five months, Šikšnys struggled through submissions, rejections, revisions, and editorial delays before finally seeing his work published online in the *Proceedings of the National Academy of Sciences* in September 2012.[62] Doudna and Charpentier found a smoother path to publication. They wrote up and submitted their results on June 8, 2012. Less than three weeks later, their paper appeared online in the prestigious journal *Science*. As Doudna notes, "Nothing after that would ever be the same—not for me, not for my collaborators, and not for the field of biology."[63]

Although Šikšnys's findings were originally submitted two months before Doudna and Charpentier's, his work was not published until well after their work had begun making serious waves in the scientific community. In science, such inequity between paths to publication and recognition of priority is more rule than exception. It becomes a fruitless exercise when studying priority disputes to discern who actually made a discovery. Science simply does not work that way. There is no objectively quantifiable moment of scientific discovery.[64] No singular event in space and time can be recorded or attributed to an individual. Scientific discovery is a fundamentally messy, communal process full of social and interpretive biases nearly impossible to unravel.[65]

A host of reasons account for different paths to publication, starting with the fact that each article submission is unique. Scientific journals have editors (and at times editorial advisory boards) who judge the viability and fit of each submitted article even before it goes out for peer review. This filtering system is necessary to manage the high volume of submissions to top-level journals. Although the articles submitted by Šikšnys and the Doudna-Charpentier team may have both analyzed the CRISPR-Cas9 complex, these were two independent research programs elucidating their topic in distinct ways. The content and framing of their discoveries cannot be compared one-to-one. Besides, the two teams submitted their articles to different journals under very different circumstances.

Though it may seem easy to point out glaring inequities in scientific publishing, comparing these submissions is no trivial exercise. Beyond differences in content, contrasting levels of social capital between the two teams clearly affected this case of publishing asymmetry. In 2012, Doudna was an internationally renowned biochemist who had already established herself as a leading expert in the "structure and biological functions of ribonucleic acids (RNAs)." She had deep roots in the international scientific community, particularly in the U.S., where she was a member of the National Academy of Sciences, the American Academy of Arts and Sciences, and had held professorships at both Yale University and UC Berkeley.[66] Finally, and perhaps most relevant to the quick publication of her CRISPR results, Doudna was a member of the board of reviewing editors at *Science*.[67] This fact does not in itself indicate wrongdoing, but it opens a window on the uneven landscape of academic publication.

Science clearly outlines its process for evaluating article submissions. Its website declares that it welcomes "submissions from all fields of science, and from any source."[68] Yet it also describes an extensive evaluation and selection process that ensures the journal continues to publish articles that are "influential" and "substantially advance scientific understanding."[69] The selection process is rigorous, entailing several stages of manuscript review. *Science* relies first on staff editors with expertise in the particular field of study to wade through all submissions. Next, selected manuscripts are sent to members of the board of reviewing editors, where they are "rated for suitability." Next comes an "in-depth" peer review in which outside readers are asked to review and comment on the manuscript. After the staff editor collects their comments, a "cross-review" occurs; all reviewers read the anonymous comments of other reviewers, then may make additional comments. Finally, when a manuscript has successfully passed all editorial stages, *Science* can accept it for publication.[70]

Even after this involved process, "Most papers are published in print and online 4 to 8 weeks after acceptance."[71] A manuscript submitted to *Science* should take at least four weeks plus the review process to arrive at publication. So it is surprising that the Doudna-Charpentier submission took only twenty days to appear online. Their expedited process does not necessarily imply impropriety or nepotism. As *Science* notes, there are loopholes in the normal editorial process for especially noteworthy manuscript submissions. When warranted, staff editors may select papers for "earlier online publication in *First Release*." *Science* also accepts "a few Research Articles for online presentation," pieces significantly longer and more prominent than regular submitted research articles. In such special cases, the editorial process seems more opaque. To access these loopholes, authors must know of them and submit a cover letter justifying special status for their manuscript.[72] Whether or not Doudna received special treatment at *Science* because of her position on its board of reviewing editors, at minimum, she had the insider knowledge to deftly navigate the publishing maze.

Here, differences in context make it impossible to fully and fairly adjudicate this priority dispute. Yet many in the sciences seem to favor the simplicity of the priority rule based on publication date.

Based on this single criterion, priority is fully determined by who is first to publish.[73] By this standard, Doudna-Charpentier would certainly receive priority. However, in addition to appearing before Šikšnys's paper, Doudna and Charpentier's publication, they have argued, was also more complete and revolutionary in articulating the CRISPR-Cas9 complex. It isolated and described the three necessary underlying mechanisms including the two mediating RNA molecules—CRISPR RNA (crRNA) and trans-activating crRNA (tracrRNA)—and the Cas9 enzyme responsible for the double-stranded cleaving. It also showed how these three mechanisms work together to precisely target and destroy invading viruses.[74]

Beyond isolating the underlying mechanisms in test tubes outside their native environment, the Doudna-Charpentier team engineered a critical enhancement to the CRISPR-Cas9 immune system. They fully programed and fused the two key RNA molecules (crRNA and tracrRNA) into one single-guide RNA (sgRNA) molecule that would work seamlessly with the Cas9 enzyme. As one commentator noted, "If there was an initial engineering feat which transformed Crispr into a 'technology,' it was the development of the [single] guide RNA."[75]

By 2015, awards and accolades were showered on Doudna and Charpentier for their pioneering CRISPR-Cas9 research. They received the prestigious 2015 Breakthrough Prize in Life Sciences, each taking home millions of dollars.[76] At the time, it seemed almost inevitable that one day they would be immortalized in the scientific pantheon with Nobel Prizes. Meanwhile, other scientists like Šikšnys were being filtered out of congealing myth-historical CRISPR narratives.[77] So Lander's "Heroes" myth-history, published in 2016, may be viewed as an attempt to reset recognition for the CRISPR development.

One clear beneficiary of the controversy was Šikšnys, who began receiving more attention and recognition for his work on CRISPR. In contrast to the Breakthrough Prize of 2015, in fall 2018, the prestigious Kavli Prize was awarded to Doudna, Charpentier, and Šikšnys for "inventing CRISPR-Cas9" and "causing a revolution in biology, agriculture, and medicine."[78] There were no qualifiers. Regardless of the publishing inequities of 2012, all three received equal status as co-inventors. Unfortunately for Šikšnys, this trend did not hold in 2020. In announcing their decision on the 2020 Nobel Prize in Chemistry, the Royal Swedish Academy of Sciences awarded it to Charpentier and Doudna for their "epoch-making" CRISPR work without mention of Šikšnys. Myth-histories will continue to be written about the invention of CRISPR-Cas9 as a gene-editing tool, but one thing is certain: Lander's early intervention has already affected the legacies of his heroes.

Lander's Lessons Learned

Although Lander begins his piece by claiming interest in a deeper understanding of the scientific ecosystem, his narrative fails to achieve this. His conclusion claims to elucidate the "lessons about the human ecosystem that produces scientific advances, with relevance to funding agencies, the general public, and aspiring researchers."[79] Far from thoughtfully drawing lessons from careful analysis of a scientific ecosystem, Lander quickly lists several points that parrot standard tropes from idealized myth-histories. He cites as important lessons from his CRISPR study the unpredictable origins of scientific advancements and the fact that many innovative ideas come from young scientists working on the community's margins. To his credit, Lander does acknowledge that "scientific breakthroughs are rarely eureka moments. They are typically ensemble acts, played out over a decade or more, in which the cast becomes part of something greater than what any one of them could do alone."[80] However, when we recall Lander's recentered and distorted narrative—essentially a hagiography of his colleague Feng Zhang—it is difficult not to be aware of the irony in his statement.

Unmistakably absent from Lander's list of lessons are critical points in any analysis of a scientific ecosystem: a network analysis that lays bare underlying social dynamics and topographies of power in the scientific community; commentary on the ongoing patent disputes that inform past scientific legacies and future research programs; a challenge to scientists to ponder ethical questions related to their work; questioning the role of scientific narratives in public misconceptions of science.

Ultimately, in filtering out unwanted noise, Lander's narrative regresses to the mean of many scientist-storytellers: a myth-history. We might be satisfied to simply categorize his narrative as a myth-history. After all, it fits nicely into Leon Lederman's template from Chapter 1. In Lederman's use of myth-history, we saw that his primary goal was to convince Congress, and the voting public, to secure funding that had been earmarked for building a multi-billion-dollar particle accelerator in Texas. Lederman unapologetically wrote a myth-history that was not historically factual because it served a clear rhetorical purpose. Lander seems to borrow from Lederman's playbook: he sees the telling of history as a means to an end, not as the end itself, appearing content to sacrifice historical accuracy for rhetorical power.

Embedded in the political rhetoric of Lander's myth-history is his attempt to downplay the accomplishments of pioneering female scientists Jennifer Doudna and Emmanuelle Charpentier. Although it's unclear that he has misogynist intent, the effect is clear. As James Watson did in his celebrated 1968 book *The Double Helix*, Lander constructs a historical narrative elevating male scientists while moving rival female scientists to the margins. Much has been written about the erasure of women from the myth-histories of science. Whether or not one believes this erasure is intentional, the lasting effects on young students of science and the public at large are troubling.

Statistics show the gender gap in science-related fields: "Although women fill close to half of all jobs in the U.S. economy, they hold less than 25 percent of STEM jobs."[81] In fact, "despite decades of research documenting explanations for this phenomenon," it remains a persistent, basic problem for the scientific community.[82] In countless studies trying to understand and propose fixes for this gender gap, a multitude of factors sustain it. A lack of self-confidence and a sense of not belonging are among the most plausible explanations for the dearth of women entering science.[83] Ensuring more female role models within STEM classrooms and the workforce is one popular proposed solution.[84]

Recent research insists that we should encourage and support women in science with an eye toward gender parity, not just because that is the ethical course, but because it would lead to greater creativity and more effective problem-solving.[85] With this in mind, you might assume the scientific community is doing everything possible to achieve gender parity. Yet negative gender stereotypes and other barriers persist. Whether perceived or structural, these barriers to entry are extremely difficult to overcome.

This is especially true when most important scientific myth-histories have been written by prominent white men, like Lander, who are insensitive to the effects of marginalizing women scientists in their narratives. To her credit, Doudna did not let Lander dictate the narrative by remaining silent. She responded to Lander's myth-history with one of her own.

Doudna Responds

On June 13, 2017, Jennifer Doudna and her colleague Samuel Sternberg released a long-form myth-history of their own. There is no mention of Lander's narrative in their book *A Crack in Creation: Gene Editing and the Unthinkable Power to Control Evolution,* but it is hard to ignore their repeated refutations of his account. Though both narratives must be considered myth-histories, by comparing their context and content we can appreciate the variety and power of this narrative form. The contrast between *A Crack in Creation* and Lander's myth-history in both tone and substance could not be more stark.

The book opens with a curious admission in the Prologue that a reader might gloss over if unfamiliar with the controversy

around Lander's CRISPR myth-history: "This book is not intended
to be a rigorous history of CRISPR or an exhaustive chronology
of gene editing's early development." The authors acknowledge
that their story is firmly rooted in their own perspective, describing how their "work dovetailed with others' research."[86] Rather
than claim, as Lander did, that they had done rigorous archival
research to reconstruct a representative picture of the "miraculous ecosystem" of science, Doudna and Sternberg lead by highlighting the limitations and possible bias of their own story.
Although clearly writing another myth-history, in direct contrast to Lander, the authors admit their own bias.

Doudna and Sternberg also take a dramatically different stance
from Lander in characterizing "heroes." Instead of highlighting a
"dozen or so scientists" as *the* heroes of CRISPR, Doudna and
Sternberg "humbly acknowledge the countless scientists who
have played crucial and invaluable roles in the study of CRISPR
and gene editing, and we apologize to the many colleagues whose
work we didn't have space to mention."[87] As we have seen, Lander's
claim of inclusivity seemed to distract from his ultimate goal of
intentionally centering Zhang and supporting the Broad Institute's
position in the patent interference case. Doudna and Sternberg's
narrative begins by avoiding any use of—or even reference to—
"heroes." Their text strikes a note of authenticity by claiming
humility in the face of the "countless" scientists who have worked
on CRISPR, and they apologize for omissions in their story.

From their Introduction, the authors seem concerned about
being inclusive in crediting and acknowledging contributions in
the development of CRISPR. In noting the authors' reflective historiography, it may be telling that Doudna's mother was trained
as a historian and taught history at a college in Hawaii.[88] So why
should we consider Doudna and Sternberg's book a myth-history?
Not all myth-histories are created the same. We cannot, and
should not, arbitrarily lump all these narratives into one category
and judge them Whiggish. Each one has a particular intent and
context, and should be understood that way. In Doudna and

Sternberg's book, we should recognize the elements that make it a myth-history while appreciating how much it differs from Lander's myth-historical narrative.

Beyond the authors' admission that their story is not intended to be a "rigorous history of CRISPR," they rely on standard myth-historical tropes to present the development of their work. Doudna and Sternberg refer to the history of science as a "long march," leaving the reader to imagine the typical myth-historical trope of a directed, inevitable progress narrative.[89] Throughout the first part of their book, the authors repeatedly purify their story of CRISPR, purging it of any meaningful reflection on scientific controversies and historical contingencies, ensuring that whatever stories are told do not tarnish the overall reputation of the scientific ideal. They consistently avoid discussing the social dynamics at the core of scientific practice and omit serious reflection on issues such as priority disputes, nationalism, differing levels of social capital, bias in awarding grants, inherent inequities in the peer-review system, and repercussions of coupling scientific innovation directly to patents. The history they present describes a science that appears objective, known, inevitable. The practice needed to accumulate purified scientific knowledge is presented as communal, collaborative, and iterative—all inevitable outcomes of a hyper-rationalized, idealized scientific method. In that sense, Doudna and Sternberg, like Lander, project and preserve an image of science as a miraculous ecosystem.

In the first part of *A Crack in Creation,* the authors establish their version of the development of CRISPR, implicitly correcting much of what they saw as distortion in the Lander piece. The second half of the book projects future implications of this revolutionary technology. Here, they take up bioethical concerns and other controversial issues absent from Part 1. The book's structure reinforces the mythological ideal that the history of science has rigorously, cumulatively produced the bedrock of objective knowledge upon which we stand today. From the privileged present we can look back and assuredly judge all past science while confidently seeing the path ahead.

In sum, their book pays homage to the celebrated scientific ideal in which we stand on the shoulders of giants. If this foundation were at all shaky, the best path forward would not be clear, so myth-historians like Doudna and Sternberg take great care to present an assured, purified history of their subject. Their book can be understood as a manifestation of the tension inherent in Einstein's dual characterization of science. Past science is "existing and complete," while science in the making "is as subjective and psychologically conditioned as any other branch of human endeavor."[90] The following section exemplifies this tension Einstein raised between a completed scientific past and a subjective, humanly fallible present.

The Case of Rosalind Franklin

Doudna traces her interests in biochemistry to the moment her father, a professor of literature at the University of Hawaii, gave her at the age of twelve a "tattered copy of James Watson's *The Double Helix.*" Doudna was thrilled to find that the book was essentially a detective novel:

> Reading Watson's account of the incredible academic collaboration with Francis Crick that had enabled them—using crucial data collected by Rosalind Franklin—to discover this simple and beautiful molecular structure, I felt the first tugs of interest that would eventually guide me onto a similar path.[91]

Unfortunately, except for the interjection momentarily citing Rosalind Franklin and her "crucial data," Doudna never again mentions the pioneering X-ray crystallographer (see Figure 3.10). Ironically, Watson derided and distorted Franklin and her work in this bestselling memoir that Doudna reports as her inspiration to become a scientist. *The Double Helix* is both highly celebrated and controversial. Although it has been prominent on lists of most

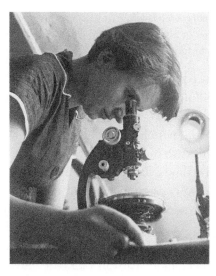

Figure 3.10 Rosalind Franklin at work, circa 1955. (Photograph by MRC Laboratory of Molecular Biology. This file was derived from: Rosalind Franklin.jpg:, CC BY-SA 4.0. https://commons.wikimedia.org/w/index .php?curid=77075413.)

influential nonfiction books, many found it incendiary even before publication in 1968.[92] As Watson was completing his manuscript in fall 1966, Francis Crick, Maurice Wilkins, and others protested so furiously that Harvard University Press was forced to abandon the project.[93]

Challenges to the book primarily targeted Watson's raw, callous portrayal of the people and events he experienced in the early 1950s. No one doubted that this was a genuine memoir presenting Watson's perspective on the discovery of the structure of DNA, but he seemed to make no effort to modulate his perspective, fact-check, or contextualize his story. In response to critics, Watson unapologetically complained that Crick and others had missed the intent of his book:

> Your argument that my book contains far too much gossip and not enough intellectual comments misses entirely what I have tried to do. I never intended to produce a technical volume aimed

only at historians of science.... Someday, perhaps you or Maurice,
but if not some graduate student in search of a PhD will write a
balanced scholarly historical work.[94]

Multiple "balanced scholarly historical" works indeed emerged,
creating a powerful potential counterpoint to Watson's story.[95] Yet
his myth-history persists, and scholarly corrections have struggled to change the popular narrative. Written in a sensational
style that continues to captivate, *The Double Helix* sold millions of
copies and has become a classic work of nonfiction. Although the
more scholarly counter-narratives validly criticize Watson's book,
they suffer from competition with a powerful, sensationally written myth-history. Watson's narrative is persistently derided for his
demeaning, misogynistic portrayal of women, especially his treatment of Rosalind Franklin. Early in *The Double Helix*, Watson introduces Franklin by focusing on her appearance:

> By choice she did not emphasize her feminine qualities. Though
> her features were strong, she was not unattractive and might
> have been quite stunning had she taken even a mild interest in
> clothes. This she did not. There was never lipstick to contrast
> with her straight black hair, while at the age of thirty-one her
> dresses showed all the imagination of English blue-stocking
> adolescents.[96]

Describing Franklin's difficult working relationship with Maurice
Wilkins, Watson continues to demean her:

> Clearly Rosy had to go or be put in her place. The former was
> obviously preferable because, given her belligerent moods, it
> would be very difficult for Maurice to maintain a dominant pos-
> ition that would allow him to think unhindered about DNA.[97]

Much has been written about how Rosalind Franklin was
wronged in both life and legacy by the misogynistic, anti-Semitic
environment of science in the early 1950s. Corrective histories
have worked to dissect Watson's myth-historical memoir.
Scholars have alleged that Franklin's work was blatantly stolen
and used to cement the prestige and legacy of Watson, Crick, and

Wilkins. These studies find her contributions have been mostly erased or minimized by the scientific community. Franklin should receive significantly more credit for unraveling the structure of DNA than she generally does. Yet Watson's myth-history has such a powerful grip on scientists' imagination that even someone like Doudna, who is clearly responding to Lander's egregious myth-historical erasure of her own contributions to the CRISPR revolution, recalls nostalgically how she was inspired by *The Double Helix*. Without hinting at the controversy around Franklin's erasure from the history of science, Doudna and Sternberg matter-of-factly refer to the discovery of the molecular structure of DNA as "Watson and Crick's discovery."[98]

This illustrates three important points. First, it shows the power of myth-histories to persist in what Lederman described as the scientific community's "collective consciousness." Entrenched myth-histories have a form of "truth" associated with their social imaginaries that is difficult to uproot and overturn, even when clear evidence disputes that "truth."[99] In perpetuating the classical myth-historical tropes of ideal science in their book, Doudna and Sternberg passively (and, to my mind, ironically) reinforce Franklin's erasure. Not to connect the treatment of Franklin by Watson to that of Doudna by Lander seems a lost opportunity. If the authors had used the Franklin case as a way to address larger issues of women's erasure from the history of science, they would have compellingly underscored the potentially corrosive effects of narratives like Lander's.

The second point Doudna and Sternberg's omission of Franklin illustrates is the power of myth-historians' commitment to the larger cause of perpetuating a purified, idealized image of science. Even if mentioning a relevant scientific controversy might help their immediate argument, myth-historians seem to weigh the benefits against the greater harm of failing to support the ideal of science.

Third, although *A Crack in Creation* perpetuates the myth-historical erasure of Franklin by celebrating Watson's account, it also points out that *The Double Helix* has inspired scores of young scientists. This is a critical point, for, as we saw in Chapter 1, myth-history may inspire in a way scholarly history does not. This form of historical narrative has an undeniable benefit that, according to Lederman and other myth-historians, should be acknowledged. From the outset, Doudna and Sternberg clearly prefer an inspirational narrative over one that is historically rigorous.

Doudna's Corrective Myth-History

Doudna and Sternberg readily acknowledge that they did not intend to write a "rigorous history" of the CRISPR revolution. What rhetorical work were they doing? Knowing what we know about the context, it is impossible to ignore Doudna and Sternberg's direct, yet unstated, corrections of Lander's "Heroes of CRISPR" piece. Lander intentionally minimized Doudna's contributions in order to place Zhang and the Broad Institute on center stage. Lander buries Doudna's work by introducing it as a momentary sidebar in which her in vitro experiments simply confirmed and extended what was already achieved. Now Doudna and Sternberg write their own myth-historical narrative, returning Doudna to the main stage.

To counter Lander's characterization of her work as a tangential, flash-in-the-pan moment in the development of CRISPR, Doudna traces her interest and research into this subject back to 2006.[100] The authors craft their historical narrative, including all the major moments that appear in Lander's timeline, while showing how Doudna's lab directly or indirectly contributed to these achievements. Resisting Lander's depiction, Doudna emphasizes her own research group's role in the fledgling, interconnected web of labs working simultaneously all over the world before 2012 to understand the CRISPR system:

> The CRISPR project epitomized this aspect of science: a few researchers around the world weaving the fabric of what would eventually become the vast tapestry of the CRISPR field, with all of its applications and implications. And in the quest to learn

more about CRISPR, our little team and many others were driven by the same sense of collaboration, of shared excitement and curiosity, that had pulled me into the world of scientific research in the first place.[101]

She carves out a space for herself in the priority dispute and explicitly emphasizes the ideal trope of science as open and collaborative, not controversially competitive.

Doudna is especially eager to correct Lander's distortion of the importance of her contributions. Lander had only included her in the in vitro chapter of his narrative corresponding to 2012 on his timeline, but Doudna's lab had been publishing results of their CRISPR experiments since 2010. Far from simply confirming the community's understanding of CRISPR in 2012 as an adaptive immune system, Doudna's lab had been working for some time trying to understand exactly how this system functions. She acknowledges that before her lab's in vitro work, biologists broadly understood CRISPR as an adaptive immune system, but "nobody really knew *how* this all worked." Doudna's team "wondered how the various molecules that made up such a system acted together to destroy viral DNA and what exactly happened during the targeting and destruction phases of the immune response." Doudna and Sternberg point out that learning exactly how this system worked "would require us to move beyond genetics research and take a more biochemical approach—one that would allow us to isolate the component molecules and study their behavior."[102] Without mentioning Lander or his myth-history, the authors fill gaps in Lander's context and address distortions.

Although they make many specific corrections to Lander's portrayal, one will highlight the rhetorical recentering Doudna and Sternberg accomplish throughout *A Crack in Creation*. As we saw earlier, Lander emphasizes the work of Virginijus Šikšnys. It diffuses credit for the 2012 in vitro work to multiple groups, implicitly questioning the priority claims of Doudna and Charpentier. Presumably Lander used this framing to dilute their achievements and complicate their patent and priority claims. Doudna and Sternberg are careful to credit Šikšnys with important contributions while undermining Lander's challenge to their priority. Doudna and Sternberg acknowledge that Šikšnys's group published "a similar paper to ours in the fall of 2012....But they failed to uncover the crucial role of the second RNA (called tracrRNA), which we had demonstrated was an essential component of the [CRISPR-Cas9 system]."[103]

The authors clearly recognize that Šikšnys's group published a similar paper, but they highlight important differences. First, his paper appeared in the fall of 2012, months after theirs. Second, although similar, Šikšnys's paper did not recognize the "crucial role" of tracrRNA in the CRISPR-Cas9 system. As the authors explain, the tracrRNA is an essential component of the CRISPR-Cas9 system that enables it to successfully and precisely target and cleave DNA. Last, Doudna and Charpentier's paper went "a step further and reengineered the RNA guide, made up of two separate RNA molecules in bacteria (CRISPR RNA and tracrRNA), into a single-guide RNA molecule that still enabled Cas9 to find and cut a particular DNA sequence."[104] All these reasons are implied as clear justifications for giving priority to Doudna and Charpentier over Šikšnys.

One point is left out of Doudna and Sternberg's treatment of Šikšnys and his work that appears in Lander's account: the difficulty Šikšnys faced in getting his paper published through the peer-review process. As we have seen, Šikšnys submitted his original paper weeks before Doudna and Charpentier. However, while theirs was fast-tracked and published in *Science* in under three weeks, his was initially rejected outright, then languished in review for five months. Doudna and Sternberg steer clear of this iceberg. They fail to mention Šikšnys's struggles to publish or their own good fortune in capitalizing on Doudna's relationship with the editorial board of *Science*. As a result, these omissions work to support Doudna and Charpentier's claims of priority and to maintain the veneer of an ideal science in which controversy, social capital, bias, and power play no part.

All the authors' corrections to Lander's narrative are done matter-of-factly. They call no attention to the priority dispute or patent controversy. There is no mention of Lander's myth-history anywhere in Doudna and Sternberg's text, and the patent controversy is mentioned just once, in a brief comment on events of 2012 and 2013:

> It was a heady time. I was elated that the work published with Emmanuelle the preceding summer had inspired others to pursue a line of experimentation similar to our own. Only later would the contents and publication dates of these papers be dissected to support arguments in a dispute over CRISPR patent rights, a disheartening twist to what had begun as collegial interactions and genuine shared excitement about the implications of the research.[105]

The authors refer to the dispute as a "disheartening twist," intimating that a scientist should not normally be involved with such matters. Apparently, these activities diverge from the Mertonian scientific ideals of disinterestedness and shared communal knowledge. Yet while Doudna and Sternberg shy away from discussing the patent dispute, as we saw in our analysis of Lander's piece, their narrative clearly does rhetorical work within that dispute. The myth of the scientific ideal is so strong that, even when scientists have been harmed or unfairly treated due to the politics of power, they continue to see—or pretend to see—a clear boundary between science politics and science itself.[106]

Lander's myth-history framed the transition from understanding the CRISPR-Cas9 system in bacteria to harnessing the technology for use in human cells as a critical technological innovation. In centering his global timeline map on Boston, he named Zhang winner of the patent dispute and credited him for carrying the scientific community across the technological threshold. Doudna and Sternberg's narrative treats that transition far differently. They give much more credit to their own June 2012 paper, in which they proposed the CRISPR-Cas9 system as a gene-editing tool for human cells. Far from making this next step seem like a giant innovative leap, they treat it as an obvious, inevitable extension:

Plenty of labs before us had succeeded in transplanting proteins and RNA molecules from bacteria to human cells, and there were many molecular tools at our disposal that we could use to help CRISPR work efficiently outside its natural environment. We just had to show that it worked as expected.[107]

As noted earlier, classifying which steps in developing a technology are obvious, and which are truly innovative, is critical to any patent dispute. Here the authors lay claim to the revolutionary importance of their June 2012 paper, labeling subsequent steps by Zhang and others as obvious extensions. By grouping their own work, and that of others, as part of the push to extend the use of CRISPR-Cas9 to human cells, they normalize the process of extension, making Zhang and the Broad Institute's claims to patentable innovation seem exaggerated. Their own claim to priority becomes explicit when the authors refer to applications emerging from three biotech companies founded to commercialize the CRISPR technology:

> By the end of 2015, these three companies would raise well over half a billion dollars more for research and development of therapies to target numerous disorders, from cystic fibrosis and sickle cell disease to Duchenne muscular dystrophy and a congenital form of blindness, all using the CRISPR technology that Emmanuelle and I had first developed and described.[108]

Doudna and Sternberg used *A Crack in Creation* to counter Lander's myth-historical narrative published in *Cell*. To do so, the authors constructed their own myth-history, a narrative with a goal far from presenting a rigorously researched history of developing a revolutionary biotechnology. There is little doubt that they sought to produce a narrative with rhetorical teeth, which might support their claims in priority and patent disputes. However, Doudna and Sternberg's myth-history is not as singularly focused as Lander's "Heroes of CRISPR." There is no flattening of discovery to a timeline that magnifies some scientists while minimizing others in the name of historical rigor. Absent is the global map that directs readers' eyes to a single laboratory.

Doudna and Sternberg did construct a narrative slanted toward their perspective and reliant on mythological tropes of ideal science, but that was stated from the outset. Unlike Lander, who claims to be exploring the ecosystem of science, Doudna and Sternberg follow through to examine the bioethical issues raised by CRISPR-Cas9. About half their book addresses these issues; although the discussion is not particularly nuanced, the authors are engaged in the larger discourse surrounding technology.

Historians of science might quibble with the characterizations presented in *A Crack in Creation*, but as long as the authors candidly acknowledge that their version of history is a myth-history, this seems less important. More interesting is to assess the impact of these myth-historical characterizations on various audiences. I applaud Doudna for reasserting her own agency against Lander's distortive narrative. It seems likely that the Royal Swedish Academy of Sciences sided with her telling of CRISPR's development over Lander's in their 2020 Nobel Prize deliberations. Although one might question Šikšnys's omission in this decision, the historic moment seems justified.[109]

In part due to Rosalind Franklin's early death in 1958 and the hostile work environment for women in science, Franklin was never able to defend herself and her work. In the intervening years the culture wars have shifted our discourse. Doudna felt empowered to push back, and many other scientists and commentators flagged Lander's narrative as misogynistic. Though I think Doudna missed an opportunity to reflect on the parallels between her erasure by Lander and Franklin's by Watson, it is difficult to ask a lone scientist to openly question and reject the myth-histories the community has long held canonical.

Exploring the Power and Impacts of CRISPR-Cas9

In the second part of their book, Doudna and Sternberg analyze the potential impact of CRISPR-Cas9. The authors reckon with the power of their invention. Beginning by showcasing what they call the "CRISPR menagerie," they list an amazing array of genetically modified organisms that already exist: mosquitoes "unable to transmit malaria"; super-muscular dogs that "make fearsome partners for police and soldiers"; hornless cows; designer pets such as micropigs; and dozens of improved breeds of livestock and crops with higher yields and other benefits.[110] Their excitement about applications of CRISPR-Cas9 shines through. They find particular promise in the use of pigs as "bioreactors" that will one day help to produce therapeutic medicines and human organs for xenotransplantation.[111]

All this is unsurprising given the billions of dollars invested in research and development of CRISPR-Cas9 applications. A generous portion of this money has funded Caribou Biosciences, a company cofounded by Doudna.[112] With multiple patents and licensing agreements in place, Caribou is a major player, pursuing R&D in livestock, agriculture, industrial biology, and drug development.[113] To her credit, Doudna does not conceal her involvement in commercializing CRISPR-Cas9 technology or having a financial interest in the patent dispute.[114] One understands her reluctance to unpack the patent controversy while the case remains in litigation, yet, again, it's a lost opportunity to look at social and economic forces that myth-histories rarely examine.

Although Doudna and Sternberg's historical narrative remains true to the myth-historical ideals of science by omitting discussion of historical controversies and patent disputes, what really sets their book apart from myth-histories like Lander's is their willingness to grapple with bioethical questions about where CRISPR-Cas9 technology is headed. They clearly wanted *A Crack in Creation* to do more than correct Lander's narrative, for in the second half they confront the tension between the exciting prospects of the technology and the "unsettling" reality that it has been relatively unchecked, even as its power and reach continue to grow.[115] Although the authors approve of much of the "menagerie" already produced using CRISPR-Cas9 and other gene-editing technologies, they distinguish between positive applications and those that may be "frivolous, whimsical, or even downright dangerous."[116] Among this group the authors list the micropig, "woolly mammoths, winged lizards, and unicorns,"[117] yet they suggest no criteria for dividing the useful from the possibly harmful.

Doudna and Sternberg seem especially bullish on gene-editing applications associated with livestock, crops, and drugs. They argue that the GMO controversy that crippled potentially beneficial innovations like the fast-growing AquaAdvantage salmon or the environmentally friendly Enviropig are casualties of a disturbing "disjunction between scientific consensus and public opinion," due, in part, to a "breakdown in communication."[118] From their perspective, GMO products have overwhelmingly been shown to be safe and beneficial, but they have catastrophic reputations rooted in misunderstanding and miscommunication. The authors see CRISPR-Cas9 as an opportunity to change the discourse around GMO. Instead of engineering an organism's genetic code by splicing in foreign DNA, à la Frankenstein, CRISPR-Cas9 can alter an organism's own DNA by precisely tweaking a specific sequence of letters. According to Doudna and Sternberg, along the blurred boundary between "natural" and "unnatural," this type of modification seems more responsible than what humanity has been doing for thousands of years. They quote Luther Burbank, a "pioneering agriculturist," who, at the dawn of the twentieth century, remarked that all living species are dynamic and moldable like "clay in the hands of the potter or colors on the artist's canvas, [they] can readily be molded into more beautiful forms and colors than any painter or sculptor can ever hope to bring forth."[119]

The single line Doudna and Sternberg refuse to cross—their "Rubicon"—is the line between somatic and germline editing in human beings.[120] Though germline editing in other species has been beneficial, "its use in humans poses significant safety and ethical challenges."[121] They pose reasonable bioethical questions: Should we be "tweaking the *Homo sapiens* gene pool in a way that cannot be easily reversed? Are we as a species prepared to assume control of—and responsibility for—our own evolution?"[122] Although these questions seem very reasonable, why do the authors feel so confident that we are prepared to assume control and responsibility for the evolution of *other* species? Why should the line between *Homo sapiens* and other species—animal or plant—be our Rubicon? Why not consider and anticipate that the world may need a more conservative threshold for genetic modification?

Although Doudna and Sternberg raise important bioethical concerns about the future of gene editing, their "techno-optimism" and commitment to the myth-historical ideals of scientific progress lead them to conclude that, thanks to technology like CRISPR-Cas9, humanity is "already supplanting the deaf, dumb, and blind system that has shaped genetic material on our planet for eons and replacing it with a conscious, intentional system of human-directed evolution."[123] To their credit, they realize that in order for this to be done responsibly, a more "global, public, and inclusive conversation" about the bioethical implications of these innovations should occur.[124] This conversation would involve a new generation of scientists that "embrace an ethos of 'discussion without dictation' in deciding how science and technology should be deployed."[125] Ultimately, they hope such discussions will "break down the walls that have previously kept science and the public apart and that have encouraged distrust and ignorance to spread unchecked."[126]

Conclusion

I wholly agree with the authors on this point. We need much more transparency in our discourse about science and its applications. Especially when compared to Lander's narrative, *A Crack in Creation* moves in the right direction. Although still relying on myth-historical tropes, it corrects much of Lander's distortion, and begins to engage Einstein's notion of science as somewhat "subjective and psychologically conditioned." A point of caution here. If we continue to sanitize our myth-historical narratives to protect the image of science at all costs, we lose opportunities to reflect and learn from our scientific past, or to be properly concerned about its future. We see this in Doudna and Sternberg's failure to recognize the controversy over Watson's denigration of Franklin in *The Double Helix* and the authors' subsequent failure to chide Lander for his similar treatment of Doudna.

Another example of cautious sanitization that could limit the open discourse the authors hope to generate is their treatment of eugenics. Introducing eugenics, Doudna and Sternberg call it "a fallacious early-twentieth-century set of beliefs and practices that have since been thoroughly repudiated by mainstream science."[127] This is a common myth-historical strategy of purification: looking back from our presentist perch, judging particular aspects of past science as fallacious, unscientific, or pseudoscientific. As many historians of science have shown, eugenics was actually a thriving, mainstream scientific subfield in the early part of the twentieth century. Doudna and Sternberg introduce Luther Burbank as a celebrated botanist and agricultural pioneer, but fail to mention that he was also a committed eugenicist.[128]

Exploring the Hall of Science at the 1933 World's Fair in Chicago, physicist Richard Feynman might have walked past a eugenics exhibit titled "Pedigree-Study in Man," four wall-panels forty feet long.[129] Like most visitors that year, Feynman was unaware of the controversy that would soon engulf this subfield. Eugenics was integrated into the Hall of Science as just another accepted subfield. The exhibit was planned as public education on this poorly understood cousin of genetics.

The term "eugenics" was coined by Charles Darwin's cousin, Francis Galton, in the late nineteenth century. Galton was convinced that through selective breeding practices, humanity might take control of evolution and the fate of our species. In a remark eerily close to that of Doudna and Sternberg's, Galton wrote: "What Nature does blindly, slowly, and ruthlessly, man may do providently, quickly, and kindly."[130] Far from a fallacious belief debunked a century ago, Galton's idea of our power to mold evolution reso-nates with the views of many of today's scientists. Galton's work on eugenics was seen by many of his contemporaries as mainstream, even pioneering. He was one of the first to rigorously apply statis-tical methods to social and biological problems. Analyzing sweet peas, he developed groundbreaking theories of statistical correl-ation and regression that quantified a modern theory of heredity.

Eugenicists who followed Galton developed his ideas. The sci-ence of eugenics became an accepted approach to advancing humanity's understanding of heredity, and applying that know-ledge to improving the human condition. By the time eugenics earned an exhibit in the Hall of Science, many established scien-tific journals and societies in the U.S., Canada, and Europe were dedicated to these pursuits. For example, in November 1933, the president of the Eugenics Society of Canada, William L. Hutton, declared that "Eugenics is no longer a fad. It is deserving of the most profound consideration." According to Hutton, the pur-pose of eugenics was not to breed superhumans. Rather, eugenics was humanity's answer to an existential crisis. "Nature" had a clear "plan of attack" to gain the upper hand on humanity: by flooding the human pool of heredity with low-grade characteris-tics, nature would eventually overcome the biggest threat to its existence, the high-functioning human brain. Eugenics was a way of directly countering nature's attack.[131]

When the Nazis adopted this branch of the natural sciences to rationalize genocide—indeed to do away with "inferior" citizens of their own country—scientists realized they had a serious "branding" problem. As many historical examinations of eugen-ics show, in the next decades the science was not overthrown. Eugenics by a different name continued to be studied, and public policies based on these studies continued all over the world. However, various branches of eugenics were disassociated, rebranded, and ultimately rehabilitated.[132]

Today, the science of eugenics lives on under new names like evolutionary psychology, in vitro fertilization (IVF), and prenatal screening. Meanwhile, the invention of CRISPR-Cas9 may have given humanity the ultimate tool to realize Galton's original vision. Knowing this shadowed history, we must be alert to how the science of eugenics was hijacked by powerful social forces a century ago—and on guard against the next wave of potentially catastrophic applications of a new technology.

One need only to review current debates on the CRISPR bombshell dropped in November 2018. Chinese scientist He Jiankui, with assistance from Michael Deem, a U.S.-based physicist and bioengineer, is using CRISPR technology to remove the CCR5 gene from viable human embryos in order to make offspring "resistant to HIV, smallpox, and cholera."[133] In a YouTube video, He Jiankui announced to the world that "two beautiful little Chinese girls named Lulu and Nana came crying into the world," the world's first CRISPR-edited babies.[134] Although many prominent scientists now demand a moratorium on this application of CRISPR technology, others are trying to navigate the bioethical issues while working to perfect such gene-editing techniques in nonviable human embryos.[135]

"The Heroes of CRISPR" and *A Crack in Creation* are myth-histories we should not dismiss. Each has a rhetorical agenda and distorts the historical development of CRISPR-Cas9, with far-reaching effects. Historians and STS scholars must study these narratives critically to understand how they affect the scientific community's and the public's understanding of science. We should take advantage of scientists' willingness to engage in inclusive discourse to bridge the chasm between Einstein's objective science of the past and subjective science in the making. As long as scientists cling to the idea that these human activities are cleanly divided, their myth-histories will continue to purify the past, control the present, and project an overly-optimistic future.

4

Echoes of Gravitational Waves

Making Heroes and Anti-heroes

Do I contradict myself?
Very well then I contradict myself,
(I am large, I contain multitudes.)[1]
—WALT WHITMAN

On the Shoulders of Giants

Sir Isaac Newton is often credited with the iconic aphorism "If I have seen further, it is by standing on the shoulders of giants."[2] While he wrote this in a private letter to Robert Hooke dated February 5, 1676, the saying can be traced at least to the twelfth century.[3] Newton's metaphor points to a core principle of many scientist-storytellers. They tend to construct their "giants" from a curated selection of past scientific activity that has been purified, translated, and rolled into a myth-historical framework that serves their rhetorical objectives and reinforces essentialist ideals of science. Myth-historical grand narratives are constructed by linking hagiographic snapshots of heroic scientists into a chain of inevitable progress (see Figure 4.1). As we have seen, readers of these narratives do not gain a realistic portrayal of past scientific practice; instead, they are fed portraits of science that rely on tropes of idealized community norms and unrealistic notions of cumulative progress.

Although these ideal norms—including universalism, organized skepticism, disinterestedness, and communalism—are effectively

Figure 4.1 "Standing on the shoulders of giants." Myth-history is constructed out of an idealized progress narrative. Here, a scientist reaches the giant's shoulder by going up the "escalator of progress" described in Chapter 1. (Source: Illustration by Zeyu [Margaret] Liu.)

impossible to meet, science students are told that, along with the scientific method, these absolute standards must always be followed if they hope in time to sit atop the shoulders of scientific giants. As more and more myth-histories appear, they reinforce the grip of distorted hagiographies of scientific giants on the community's collective conscience. Is this the best way to present science? Do these apocryphal tales starring mythologized heroes inspire would-be scientists and the public at large? Or do such tales spread negative unintended consequences?

Myth-historical science stories resound clearly in the classroom. Unfortunately, there is little reflection on how they affect young students. Many science teachers assume that myth-histories offer positive role models that inspire students, but is this accurate? In a 2016 study, researchers found that some high school students were motivated more by struggle narratives of iconic scientists than by myth-historical hagiographies.[4] The struggle narratives carefully contextualized scientific achievements by including both personal and professional difficulties and obstacles to success. They made heroes like Albert Einstein and Marie Curie somewhat more relatable to science students.

An especially interesting finding of this study is that not all students responded equally to struggle narratives in place of heroic myth-histories. Like all scientists, students will inevitably struggle with some aspect of their scientific practice. It is telling to see how each student metabolizes the contrast between their own imperfections and the myth-historical ideals they are given. Do they internalize this dissonance as personal failure, or do they protect their ego by forging ahead in their studies, keeping dissonance at arm's length? If they internalize the dissonance, it can undermine their confidence and lessen their sense of legitimate scientific self. This outcome seems more likely for students who already struggle to maintain their confidence and self-worth. Particularly if a student already feels as though they do not belong in a STEM discipline, dissonance between their reality and the myth-historical ideal may translate into an insurmountable barrier to success.

To address these matters successfully, science teachers and practitioners alike need to be more aware of the effects their myth-histories may have on young students and their sense of belonging. Myth-histories are constructed by stripping our scientific heroes of their historical, social, and human contexts, so they may be seen as larger than life. Yet there is little reflection on how these caricatures impact students and general audiences.

As we saw in previous chapters, beyond pedagogy, myth-historical impacts are also seen in research agendas of scientific communities, funding opportunities, and in the broader public discourse about science. In this chapter we look at a reconstruction of history to celebrate the first-ever detection of gravitational waves. In the winter of 2016, in a full-on public relations blitz, the NSF-funded Laser Interferometer Gravitational-Wave Observatory (LIGO) announced new, iconic measurements of the warping of spacetime by doing some rhetorical warping of their own.

Project leaders used historical interventions to strategically frame their accomplishments. With two important yet subtle revisionist twists, they turned history into myth-history. In one twist they told a simplified, inevitable progress narrative that played on the heroic trope of one of physics' patron saints, Albert Einstein, to highlight the importance of their discovery. In the second twist, they rehabilitated the reputation of Joseph Weber, a physicist who had died sixteen years earlier, marginalized and forgotten.

Comparing these two twists used to build a LIGO myth-history, this chapter shows that the boundary between scientific heroes and pariahs is both razor-thin and permeable. Heroes and pariahs from the past are part of our scientific imaginaries, continually constructed and reconstructed in the present. Delicate negotiations are executed within scientific communities to articulate topographies of power and their resulting dynamics.

LIGO Presser

On February 11, 2016, at the National Press Club in Washington, DC, France Córdova, Director of the National Science Foundation, stood at a podium to announce the opening of a "new observational window" that would allow humanity to "see our universe, and some of the most violent phenomena within it, in an entirely new way."[5] For centuries, astronomical observation was limited to using traditional optical telescopes. Thus, astronomers' observations reflected only what could be deduced from electromagnetic signals in the optical range. In the twentieth century, the range of signals observed was extended to a wide swath of the electromagnetic spectrum, including radio waves, X-rays, and gamma rays. With each new observational window, astronomers gained new cosmological perspectives that revealed previously unobservable dynamics and a deeper understanding of our universe. That winter morning, Córdova and representatives from the LIGO team announced to the world a completely new form of astronomy: gravitational-wave astronomy.

Córdova was joined by LIGO executive director, David Reitze; the spokesperson for the LIGO Scientific Collaboration, Gabriela González; and two of the three cofounders of LIGO—Rainer Weiss and Kip Thorne (see Figure 4.2). After four minutes of introductory remarks, Reitze stepped to the microphone and made the official announcement: "We did it!" The NSF and LIGO representatives basked in celebratory glory as thunderous applause rang through the hall.[6] The LIGO team had finally achieved what seemed impossible to Einstein a century earlier, direct observation of gravitational waves.

Like ripples on a pond, gravitational waves radiate out from violent cosmological disturbances. Unlike water ripples, gravitational waves are oscillations in the very fabric of spacetime (see

Figure 4.2 The LIGO team at the press conference on February 11, 2016, in Washington, DC. Left to right: France Córdova (standing) David Reitze, Gabriela González, Rainer Weiss, and Kip Thorne. (Source: Photograph by National Science Foundation; Researchers announce LIGO detection of gravitational waves, public domain. https://commons .wikimedia.org/w/index.php?curid=65646346.)

Figure 4.3 Like waves on a pond, this illustration shows gravitational waves as ripples in spacetime. In this case, they are radiating out from two in-spiraling black holes. These waves become fainter the farther they get from their origin. (Source: Courtesy Caltech/MIT/LIGO Laboratory.)

Figure 4.3).[7] A passing wave will alternatively stretch and compress spacetime and everything within it. Near the source of cosmological disturbances, these spacetime distortions can result in massive gravitational waves. However, due to the sources' cosmically large distances from Earth, by the time the gravitational waves reach the twin LIGO detectors in Hanford, Washington, and Livingston, Louisiana, they are imperceptible ripples in spacetime on the order of one ten-thousandth the diameter of a proton.[8] That is why, for much of the century since they were first postulated, these oscillations seemed well beyond the reach of human observation, hopelessly undetectable.

As Córdova noted in her opening remarks, the NSF has been funding the basic science upon which LIGO is based since the mid-1970s and the LIGO project itself since 1992.[9] The LIGO detectors were designed to be the most sensitive laser interferometers on the planet.[10] These L-shaped antennas are essentially 4 km-long rulers that use the interference of lasers to detect the slightest distortion in the fabric of spacetime (see Figure 4.4). They can be

Figure 4.4 Photo of the LIGO laboratory detector site near Hanford, Washington. It's a massive L-shaped interferometer with 4-km-long arms. A second twin detector is located near Livingston, Louisiana. They are extremely precise, able to measure disturbances at the scale of 1/10,000th the diameter of a proton! To achieve this, they are vacuum-sealed and isolated from all immediate environmental effects. (Source: Courtesy Caltech/MIT/LIGO Laboratory.)

thought of as the most precise measuring devices ever invented. In reinforcing this point, Reitze claimed that LIGO is so precise that if it could be used to measure the distance between the sun and its nearest neighboring star, over three light years away, it would do so to within the width of a human hair![11] (It's important to note that LIGO cannot currently be used in this hypothetical way.)

In the case of the first detection, the disturbance was the result of two black holes merging into one larger one, approximately 1.3 billion years ago. Computer simulations based on solutions of Einstein's field equations, and displayed at the news conference, show how this dynamically violent cosmological event would have evolved. Two in-spiraling black holes, each with a mass approximately thirty times that of the sun, packed into a space roughly the size of the New York metropolitan area, sped toward each other at half the speed of light. As they coalesced, some of their combined mass was given off as energy in the form of gravitational waves radiating out in all directions. Just at the moment of merger, the frequency of these waves would have sharply increased, causing them to bunch up. It then took that burst of emitted gravitational waves, traveling at the speed of light, 1.3 billion years to reach the LIGO detectors on Earth.[12]

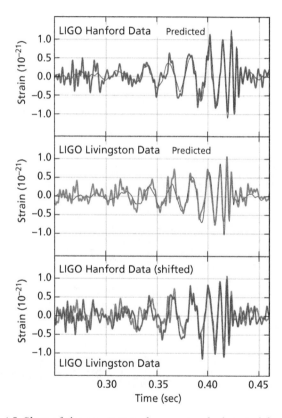

Figure 4.5 Plots of the gravitational wave signals detected first at the Livingston observatory and then 7 milliseconds later at the Hanford site. The top two are the actual data from the detectors overlaid on the predicted wave forms. The third plot compares the data from the two sites. (Source: Courtesy Caltech/MIT/LIGO Laboratory.)

On September 14, 2015, as most scientists working at the two LIGO sites in the U.S. were still asleep, a postdoctoral student named Marco Drago was in his office at the Max Planck Institute in Hanover, Germany, routinely monitoring data feeds from the two detectors. At 11:50:45 a.m. local time, an email alert notified Drago that the LIGO detectors had registered an event.[13] The LIGO site in Louisiana had detected an anomalous waveform, and then, seven milliseconds later, the second detector—3,000 km away in Washington state—had recorded a match (see Figure 4.5). After five months of analysis to rule out any other possible explanations for the matching waveforms, the signal was certified by the LIGO team, peer reviewed by the journal *Physical Review Letters*, and announced to the world as the first-ever detection of gravitational waves. Figure 4.6 shows the results of LIGO's first observing run.

For Córdova and her colleagues, the February 2016 press conference celebrated an important scientific accomplishment. As

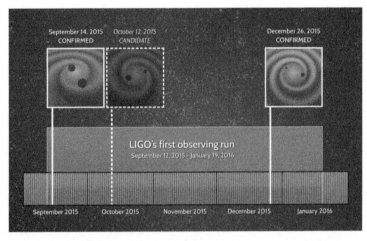

Figure 4.6 Chart showing the results from LIGO's first observing run in 2015–16. It marks the first two confirmed gravitational wave detections in September (event GW150914) and December (event GW151226) 2015. In both cases, LIGO detected the faint ripples in spacetime resulting from two black holes spiraling toward each other and merging. (Source: Courtesy Caltech/MIT/LIGO Laboratory.)

we shall see, myth-historical anecdotes were offered to context-ualize the importance of the feat. In doing so, they recast history through a presentist lens, foregrounding their current perspective and filtering out inconvenient details that might distract from the primary message, "We did it!"[14] There's nothing inherently wrong with doing this, especially given how truly momentous the occasion was. However, it's important for everyone, including scientists and their audiences, to be aware of the rhetorical choices, and distortions, made. The following discussion focuses on two interventions that helped the LIGO team deliver a coher-ent picture of scientific progress.

Myth-Historical Interventions

Rainer Weiss began the press conference by saying, "I'm going to tell you a little about history."[15] He then contextualized what LIGO was doing scientifically by anchoring it to Einstein's most revered accomplishment: the reconceptualization of gravity through his general theory of relativity (GTR), published in 1915.[16] According to Weiss, within a year of developing his famous field equations, Einstein began applying them to "the idea of gravitational waves." In the words of Weiss, "there were waves" in GTR, and Einstein "found" them. Einstein interpreted these waves, which propagated at the speed of light, as what Weiss called "strains in space" that "communicated information" about the gravitational force.[17]

After describing the size of the equipment necessary to meas-ure such minuscule strains in space, Weiss returned to the historical underpinning. Even though he admitted to having no concrete evidence to support the claim, he described Einstein as a "patent clerk" and a "practical physicist" who "in the early days of, let's say, 1916, ... probably looked very hard at doing things himself." Understanding that these waves were caused by accelerating masses, Einstein "probably put on the backs of envelopes" the first calculations for gravitational waves and their corresponding strains. "We're looking for those envelopes," Weiss said. "People don't have them, but they had to exist." He concluded by characterizing Einstein as "despondent" after realizing that, even by accelerating the largest masses then known—a system of "binary stars"—the resulting gravitational waves would be too weak to detect.[18]

Weiss's myth-historical intervention is a common trope that infuses most gravitational wave narratives. In his retelling, Einstein single-handedly developed GTR in 1915. He saw that gravitational waves must exist, but concluded that they would never be observed due to their small amplitudes. This narrative is historically problematic but has rhetorical value. As myth-histories are wont to do, the story filters out much detail that some might see as irrelevant noise of daily life. In this case, the details relate to preexisting gravitational wave theories and Einstein's ambivalence about his own work on gravitational waves. The details illustrate the communal and social dimensions of science, as well as the difficulty and uncertainty of being on the edge of scientific innovation. Before taking a closer look at these filtered details, let us return to the press conference to hear what other myth-histories were invoked that day.

Weiss was followed at the podium by one of the LIGO project cofounders, Kip Thorne. Thorne began by claiming that

> LIGO has been a half-century quest. It arose in part in the 1960s from pioneering work by Joseph Weber at the University of Maryland, and it arose from interferometer R&D in the 1970s and '80s at Caltech, MIT, in Scotland, and in Germany.[19]

Thorne went on to explain that the facilities of LIGO were built in the 1990s, operationalized in the early 2000s, and upgraded with advanced interferometers between 2010 and 2015. Thus, the apparently linear progress narrative began with Weber and led gradually but inevitably to the "spectacular results" celebrated today.

Next up was the director of NSF, Córdova, who again invoked Einstein: "Wow! Einstein would be beaming, wouldn't he?" She then commented on the occasion's significance for her personally. Córdova considered Thorne a "faculty mentor" during her graduate studies at Caltech in the 1970s. Then she mentioned another Caltech faculty member in attendance, Virginia Trimble, who, with Thorne, had inspired her with "stories of black holes, which seemed imaginary at the time." Trimble was at the press conference, at Córdova's invitation, because she was "both an astrophysicist and a historian of astronomy." But Trimble was also there, Córdova noted, because "her late spouse, Dr. Joseph Weber, did pioneering work in the '60s that Kip Thorne already highlighted." Córdova said, "It's really very special that at NSF's LIGO facility in Hanford, Washington, [Weber's] early instrument is on display."[20]

Finally, Córdova rightly pointed out that science is a collaborative enterprise:

> Discoveries of this magnitude do not happen overnight. They aren't made by one person working alone in a lab. They arise from the boldness and brilliance of scientists like Rai[ner] Weiss, Kip Thorne, Ron Drever, and Joe Weber at the start.[21]

In this one moment, Córdova effortlessly elevated Weber to the rank of scientific hero, alongside Nobel laureates Thorne and Weiss. However, as with Weiss's sketch of Einstein's contributions to the history of gravitational wave astronomy, both Thorne's and Córdova's invocations of Weber are more myth-historical than historical. They ignore the controversial nature of Weber's contributions to the field and his status as a scientific pariah for many decades.

The myth-historical interventions at the press conference on February 11 are echoed in the press kit released by LIGO the same day. In the press release, Reitze summed up the team's accomplishment with a tidy narrative that seems constructed to capture the public's imagination:

> Our observation of gravitational waves accomplishes an ambitious goal set out over five decades ago to directly detect this elusive phenomenon and better understand the universe, and,

fittingly, fulfills Einstein's legacy on the 100th anniversary of his general theory of relativity.[22]

This encapsulation stresses the difficulty of the accomplishment and ties it to the centennial of Einstein's GTR.

The myth-historical interventions presented in the press conference and the press kit helped form the standard narrative of wave detection. In turn, the narrative helped justify the billions of tax dollars spent on LIGO. For someone familiar with Einstein's ambivalence, however, or with Weber's controversial status, the myth-making seems crafted to cast uncertain heroes in new light. As in myth-histories we've encountered in previous chapters, the stories about wave detection have rhetorical objectives that tend to override historical accuracy. Let's take a closer look.

Will the Real Einstein Please Stand Up?

Rainer Weiss anchored LIGO's achievement to the legacy of Albert Einstein (see Figure 4.7). Einstein's work on relativity theory and quantum mechanics during the first three decades of the twentieth century makes him one of the greatest scientists who ever lived. Weiss is not the first to invoke Einstein. Whenever one can somehow couple their work to Einstein, it draws on the social capital of his legacy, helps to establish their credibility, and draws wider interest to the topic being discussed.[23]

Given the established connection between Einstein and gravitational waves, it is not surprising that the LIGO team would exploit the relationship. They all but invite us to imagine them standing on Einstein's shoulders, searchlights in hand, helping humanity see further into the unknown.

The problem is this. The name Einstein is synonymous with genius, but if we take a deeper dive into his place as a scientific and cultural icon, we see that his legacy is extremely fractured. There is no single Einstein—only a multitude of Einsteins—each refracted through myriad lenses. Einstein himself used an analogy to describe the difficulty of holistic understanding: "Nature

Figure 4.7 Albert Einstein. (Source: Photograph by Philippe Halsman © Halsman Archive. Courtesy of The Morris and Adele Bergreen Albert Einstein Collection at Vassar College.)

shows us of the lion only the tail.... [The lion] cannot reveal himself to the eye all at once because of his huge dimensions."[24] Historical understandings of Einstein are plagued with the same problem of partial perception.

Thus Einstein has been many things to many interpreters: a young revolutionary overturning established scientific truths; an older reactionary unwilling to accept all changes sparked by his revolutionary ideas; an abusive, misogynist husband and absentee father; a social justice warrior; a fervent Zionist and early advocate of nuclear weapons; a steadfast pacifist; a spiritually inclined philosopher; a pragmatic atheist; an uber-celebrity and playful icon; and yet, the loneliest of people. Like Walt Whitman, Einstein can be said to "contain multitudes." A Google search returns thousands of portraits animated by endless quips, quotes, and anecdotes. Some are the result of a dynamic, eclectic life; others the result of becoming an avatar for so many people and their particular causes.

For decades historians have attempted, and failed, to construct a singular Einstein, fully contextualized within his place and time, embracing all his idiosyncrasies and contradictions. Due to his celebrity, many Einstein artifacts are preserved in archives around the world, each casting him in a slightly different light. The goal for the historian of science is to examine individual artifacts in their proper contexts and then work to fit them into a larger, coherent representation of Einstein that preserves his humanity and historicity.[25]

Inevitably, this process of analysis and synthesis is messy and interpretive. Working with so many projections, historians must choose what artifacts to examine and how to weave them into a larger picture. Depending on the historian's intent and audience, their process will show some element of selection bias. However, professional historians, conscious of this, employ methodologies to minimize biases. Sampling various scholarly representations of Einstein might lead one to a fractured yet reasonably coherent picture of him as a complicated yet singularly real person.[26]

Myth-historians are not necessarily committed to representations of Einstein that are multidimensional or properly contextualized. In fact, many see engaging with the details of his life as yielding a noisy, unfocused image of the man. They prefer an uncomplicated projection with sharp edges that allows them to use Einstein for their particular rhetorical ends. Consequently, myth-historians tend to produce caricatures.

The most common of these is Einstein as patron saint, the very image of the ideal scientist. He is presented as the quintessential forward-thinking revolutionary who came out of nowhere to single-handedly push physics into its modern form. His theories upended the physics world to its core by revolutionizing fundamental conceptualizations of space, time, energy, and mass. These myth-historical versions of Einstein typically focus dramatically on his initial development of special relativity during his annus mirabilis of 1905, followed by his single-minded march toward the 1915 publication of his GTR. All Einstein's brilliance and hard work is eventually rewarded by Arthur Eddington's observations of the 1919 eclipse and subsequent confirmation of Einstein's predictions of the bending of starlight.

Only rarely are the contexts of innovation that fostered his contributions accurately touched on in myth-historical portrayals of Einstein. As historian of science Matthew Stanley observes in *Einstein's War*, the story of Einstein's relativity was a "deeply human one"; a "messy adventure that combined friendship, hatred, and politics."[27] The Eddington expeditions of 1919 confirmed Einstein's GTR and made him a giant. However, this confirmation failed to exemplify an ideal, dispassionate, objective science. Instead, GTR's birth is a sometimes "chaotic" story in a "tumultuous" time.[28]

Myth-historical narratives about Einstein usually omit a serious accounting of his early professional struggles to gain entry into the scientific community. True, many mention that he was a patent clerk or that early on he struggled in school, but such anecdotes are typically used as benign narrative devices to add drama. Anecdotes about struggle can frame Einstein's career in a narrative arc without revealing the underlying social bias and persistent barriers to success he faced. For myth-historians intent on projecting an image of science in line with Mertonian ideals, revealing a scientific past with xenophobia, prejudice, self-interest, and misogyny might make their audience question whether science really is impartial.

In Einstein's case, some of the struggle anecdotes might not even be true. He did struggle in certain school subjects, for example, but the common trope that he failed mathematics when he was young is part of a myth-historical imaginary. In 1935, after reading a headline that declared "Greatest Living Mathematician Failed in Mathematics," Einstein flatly denied it: "I never failed in Mathematics.... Before I was fifteen I had mastered differential and integral calculus."[29] Historical research has long supported Einstein's claim that he was mathematically precocious. His school grades and family correspondence leave no doubt that from a young age he was adept in technical subjects such as mathematics. Far from being a mathematical failure turned genius, Einstein was mathematically precocious. Whereas proper biographies tell of his reading Euclid and solving mathematical proofs at a very young age, myth-histories rarely mention such details as his voracious reading habits or his privileged upbringing.

In reinforcing scientific social imaginaries, myth-historians selectively filter out elements of social life that are incongruous with scientific ideals. Instead of acknowledging that science is a deeply human activity—thus entangled in the realities of prejudices and biases—myth-historians try to keep science pure. To do so, they sanctify the past by purging their narratives of anything that might be unpalatable by today's standards. Rarely do we encounter myth-histories that grapple with Einstein's real struggles to overcome barriers like the institutionalized anti-Semitism in academic communities of his time.[30]

Myth-historical narratives select biographical details that seem to fit with celebratory tropes. Among the deepest of these is the notion that science is universal and disinterested. As such, science relies on a strict methodology that allows all practitioners to describe the same reality. Any human being—regardless of race, gender, class, religion, political affiliation, or nationality—can apply the same methodology and arrive at the same rational conclusions. In theory, science is open to all and is indifferent to the biographical and demographic details of its practitioners. In this ideal, there is no room for bigotry or prejudice. There is no room for Jewish science versus Aryan science. There is no room for hatred or indeed any "subjective" force that might distort the fruits of scientific practice.

When anti-Semitism is discussed within the context of scientific myth-histories, it is generally focused on the rise of Nazism in Germany in the early 1930s and how this led to a forced migration of many prominent Jewish scientists, including Einstein.[31] This framing makes science the victim of vile external social forces. However, a more reflective account would recognize that this prejudice and hatred began to percolate decades before the Nazis took power. No social institution, not even science, was immune.[32] It began as hidden acts of institutional prejudice at the University of Zürich in 1909.[33] It grew into bolder, more public anti-Semitic hostility toward Einstein's "Jewish physics" after his celebrity in 1919.[34] Considering current efforts to dismantle barriers to success for underrepresented minorities in STEM fields, much can be learned from Einstein's struggle against anti-Semitism. We must be far more transparent about our legacies of exclusion and prejudice.

The Cost of Casting Einstein as a Solitary Genius

A more nuanced examination of Einstein's role in the discovery of gravitational waves will help uncover problems with the LIGO myth-historical narrative described earlier in this chapter. Contrary to Rainer Weiss's comments at the press conference, the story of gravitational waves did not begin with Einstein's calculations in 1916. A more careful analysis reveals that their prehistory dates back at least to the eighteenth century. Although clearly not a full gravitational wave theory, Pierre-Simon de Laplace's perturbation theory of orbital damping was a step in that direction. However, the biggest challenge to Einstein's legacy as sole progenitor of gravitational waves is the work of Henri Poincaré and Max Abraham. While Einstein was working to extend his special theory of relativity to a more general, non-inertial domain, Poincaré and Abraham were equally enmeshed in that line of research.[35]

Poincaré contributed to the foundations of special relativity theory long before Einstein's publications of 1905. Though Einstein is traditionally credited with plucking special relativity out of thin air, many physicists had been struggling with reconciling mechanics and electromagnetism in the decade before 1905. Among these, Lorentz had made his eponymous transformations explicit, and Poincaré had begun to question the absolute nature of simultaneity, space, and time.[36] These are all concepts that

Einstein would eventually use to construct his particular brand of special relativity. It is unclear to what extent Einstein was, in the spring of 1905, aware of Poincaré's contributions. What is certain is that studying the contexts of Einstein and his peers during the first years of the twentieth century shows that the time was ripe for someone to revolutionize conceptions of simultaneity, space, and time.

In 1908, Poincaré, like Einstein, was working to extend the special theory of relativity beyond its inertial reference frames. He recognized that the finite limit of the speed of light for any signal transfer should also restrict the gravitational force from acting instantaneously. Poincaré proposed an *"onde gravifique"* that would propagate at finite speeds. He was thus the first person to use the phrase "gravitational wave."[37]

While Einstein was working on his GTR, Max Abraham was producing his own "relativistic theory of gravitation." In this theory, published in 1912, Abraham included a full discussion of the possible existence of gravitational waves. Based on symmetry arguments about the source of the gravitational disturbance, Abraham was first to correctly argue that there was a critical difference between the character of the radiation stemming from electromagnetic and gravitational fields. He found that, unlike electromagnetic waves, "the lowest order of gravitational radiation must be quadrupole."[38] He argued against the existence of both monopole and dipole gravitational radiation, concluding that therefore gravitational waves must not exist. Abraham's arguments were correct; his conclusions were not. We now know that gravitational radiation does exist, and it's dominated by quadrupole radiation.

Some of the lacunae in recognizing other scientists' accomplishments is due to Einstein himself. He was eventually well aware of both Poincaré and Abraham's work on relativity and gravitational radiation, but chose not to recognize their contributions in his publications. Much has been written on Einstein's less than generous treatment of his peers' work. It's a practice he began in his youth and followed throughout his career. In fact, Einstein himself was aware of his sparing use of citations and considered it a fault. When discussing with an assistant whether to add another scientist's name in a literature review, Einstein once retorted, "Oh yes. Do it by all means. Already I have sinned too often in this respect."[39]

Myth-histories, however, instead of correcting such omissions, tend to reinforce them. One might think it would be in scientists' best interest to portray past science as collaborative, or at least communal—a scientific ideal, after all. Nevertheless, in practice historical accuracy is often sacrificed in favor of the solitary genius account. To elevate scientific heroes one needs to diminish the contributions of those around them. Apparently, heroes are so important to myth-history that they often eclipse the agenda of representing an idealized science.

What's the problem with Weiss's picture of Einstein doing calculations on the back of hypothetical envelopes and discovering gravitational waves in his new field equations? One consequence is that gravitational waves are presented as an immediate, logical outgrowth of Einstein's GTR. In fact, they were far from inevitable and not solely tied to the GTR. Poincaré's and Abraham's work on gravitational waves points to the contingency of the outcome that Weiss presents as inevitable. Once we know of the alternate parallel research programs, we can imagine a scenario where a robust model of gravitational waves might have eventually come out of a different theory of gravity. Even for Einstein, the notion of gravitational waves was far more contingent than Weiss made it out to be.

Einstein's Unshakable Ambivalence

Far from discovering gravitational waves in his field equations and becoming certain of their existence, Einstein's relationship to these field oscillations was much more complicated and ambivalent. According to correspondence between Einstein and the young German astronomer, Karl Schwarzschild, Einstein's initial calculations from November 1915 to February 1916 led him to believe that gravitational waves do *not* exist.[40] It would be months before he would reverse course and write his now-famous paper proposing their existence.

This about-face was sparked by the suggestion of Dutch astronomer Willem de Sitter to consider using a "linearized approximation" in solving the nonlinear relativistic field equations. In this new linear first-order approximation, the analogy between electromagnetism and gravitation became more evident. In both cases, field oscillations could be described using wave equations that yield propagating waves. Even so, for years after this first paper, Einstein struggled with how to interpret the gravitational waves, vacillating on their physical nature. Were they real or just an artifact of his choice of a coordinate system? Did they carry energy? Could they be observed?[41]

In mid-1936, Einstein formalized his ambivalence about the existence of gravitational waves. One year earlier, his paper written with Boris Podolsky and Nathan Rosen (hence, "EPR") had challenged an emerging quantum orthodoxy, asking "Can Quantum-Mechanical Description of Physical Reality Be Considered Complete?" Still unsatisfied by the orthodox interpretation of quantum theory, Einstein began looking for a unified field theory that would unveil deeper parallels between quantum treatment of electromagnetic fields and relativistic treatment of gravitational fields. Einstein turned his attention back to his earlier work on gravitational waves. He and Rosen wrote a paper in which they attempted to solve the nonlinear field equations without resorting to the linearized approximation used two decades earlier. In doing so, Einstein was convinced that "gravitational waves do not exist, though they had been assumed a certainty to the first approximation."[42]

Einstein and Rosen titled their manuscript "Do Gravitational Waves Exist?" and submitted it to the journal *Physical Review*. However, the paper was never published in *Physical Review*. Instead, when confronted with an anonymous critical referee report,

Einstein bitterly withdrew the submission. In the wake of the dustup he vowed never to publish in *Physical Review* again. Apparently Einstein took offense, in part, because he was not accustomed to being challenged by anonymous reviewers. The Einstein and Rosen gravitational wave paper eventually did appear in a less prestigious journal the following year, but by then it had been altered substantially.[43]

The story of the Einstein-Rosen gravitational wave paper is an example of why the details of human and social dynamics are important in understanding the history of science. In the intervening year, Rosen left for the Soviet Union, and Einstein and his new assistant, Leopold Infeld, consulted with colleagues at Princeton about the paper. Among the colleagues they consulted was the celebrated cosmologist Howard Percy Robertson, who was critical of the paper. In retrospect, that shouldn't be surprising since, unbeknown to Einstein, Robertson was the anonymous *Physical Review* referee who had critiqued the original paper! Hearing many of the same critiques again, Einstein was forced to reconsider his initial results. While revisiting the underlying assumptions of the original paper, he "stumbled upon" a new solution to his field equations representing cylindrical gravitational waves. The revised paper published in the *Journal of the Franklin Institute* was now altered substantially and titled "On Gravitational Waves." After it was finally published, Rosen was surprised and dismayed to find that their coauthored paper had appeared in a lower-tier journal, and no longer explicitly questioned the existence of gravitational waves.[44]

Contrary to Weiss's myth-historical anecdotes, Einstein's position on gravitational waves was not firmly established with his 1916 paper suggesting their existence. As Daniel Kennefick notes, Einstein "left a rich but ambiguous legacy" on gravitational waves. As a result, his evolving theoretical models should be put into conversation with his evolving skepticism over their existence. Whether or not Einstein died believing in real-world, energy-carrying gravitational waves is not clear from the historical record. What is clear is that, inspired by their collaboration with Einstein, two of his assistants, Rosen and Infeld, seemed inclined toward skepticism. In turn, they both helped spark debates in the field and had a "great impact on the future development of gravitational waves."[45]

Those debates would take time to develop. Just as there was an interpretational lull in quantum theory that lasted several decades (see Chapter 2), there was a significant lull in GR research. In the decade after the publication of Einstein's field equations, there was considerable research into the foundations of GR. But from the mid-1920s to the mid-1950s, GR was eclipsed by geopolitical tensions (including, of course, a world war) and other scientific pursuits. In particular, a dearth of applications that required GR forced the field onto the fringes of physics. During this period, GR was kept alive by a few mathematically minded physicists who worked in a fractured multidisciplinary field made up mostly of mathematicians and cosmologists.[46] The GR community was able to maintain itself through a "dispersed potential embedded in a variety of research traditions."[47]

In July 1955, that dispersed potential began to coalesce. The occasion was the fiftieth anniversary of Einstein's annus mirabilis and the fortieth anniversary of the publication of his relativistic field equations marking the birth of GR. Using the jubilee as a catalyst, a handful of physicists, including Nobel laureate Wolfgang Pauli, organized a conference on relativity theory to be held symbolically in Bern, Switzerland. Three months before the conference, Einstein died. Einstein's death juxtaposed with the conference galvanized the GR community to overcome "disciplinary and national divides."[48] For decades, Einstein's quest for a unified field theory had relegated the GR to a mere "stepping stone" toward that goal.[49] Now a community from the fringe resolved to take GR seriously.

The period that followed has been described as a renaissance: general relativity and gravitation became a mainstream, integrated subfield of physics.[50] The transition was not without tension. For decades, GR had been studied by mathematicians, cosmologists, and mathematically minded physicists, each with their own framework. Disciplinary convergence brought about an inevitable clash of cultures.[51] Certain key questions also returned to the foreground—in particular, gravitational waves. At the time of the Bern conference in 1955, there was still no consensus. Rosen reaffirmed his skeptical conclusion that because they do not carry energy, gravitational waves are not physically real.[52] However, over the next several years, a consensus did begin to emerge, motivated, in part, by some incisive thought experiments. The Bondi-Feynman thought experiment of 1957 was an especially important catalyst for a growing consensus about the existence of gravitational waves.[53]

Early Explorations into Gravitational Wave Detection

In January 1957, Hermann Bondi and Richard Feynman were among a distinguished group of physicists at a conference at the University of North Carolina, Chapel Hill, on the "Role of Gravitation in Physics." The origin story of this conference and its ties to the scientific-military-industrial complex is a compelling example of the contingent nature of scientific research and the critical roles human, social, and political dynamics play in articulating its practice. Throughout the 1950s, wealthy industrialists and the U.S. military were drawn to funding research in gravitation for its promise in seeding "anti-gravity" technologies that were more science fiction than science. The goal was to ensure that the U.S. remained technologically dominant as human activities eventually moved into space.

Scientists seeking funds for research had to navigate a delicate course between keeping potential patrons interested and keeping their science on solid footing. From the correspondence between organizers and patrons of the Chapel Hill conference, as well as from the proceedings, one gets the sense that the organizers were successful at steering that course. Anti-gravity was mentioned only once at the conference, and it was quickly disavowed.[54] The conference was a serious affair, attended by many leaders in relativity research. One of those was Princeton University's John Wheeler.[55]

Wheeler was one of the most eclectic physicists of the twentieth century, having gained expertise in numerous fields, including particle physics and GR, and leaving a broad legacy of contributions and accomplished graduate students. He had only recently become interested in GR research and had joined the steering committee to help organize the Chapel Hill conference.[56] Wheeler invited a large contingent of present and former graduate students from Princeton, including Richard Feynman, who was interested in generalizing field theories and seeking bridges between gravitation and quantum mechanics.[57]

Somewhat skeptical of the work being done in the field, Feynman showed up late to the conference and missed the session on gravitational waves. However, in the second-to-last session, on quantized general relativity, Feynman introduced a thought experiment to show that gravitational waves *must* carry energy.[58] His ideas were sparked by Felix Pirani's work on the physically observable properties of gravitational waves.[59] An expanded form of Feynman's thought experiment was included in the final conference proceedings.

The thought experiment goes like this: you could test for energy transfer by setting up a rigid rod with beads wrapped around it. As tidal (gravitational) waves pass through the rod, they cause the beads to move back and forth. This motion would result in a detectable transfer of energy in the form of heat due to friction. For Feynman and many others, this simple argument was enough to convince them of the existence of real, energy-carrying gravitational waves. Hermann Bondi, who had studied with Arthur Eddington in the late 1930s and was one of the leaders of the GR renaissance, had independently come up with a similar thought experiment earlier in the conference. Like Feynman, Bondi had relied on Pirani's calculations. Bondi went on to publish the first rigorous account of the thought experiment in *Nature* four months after the conference.[60]

Another participant at the Chapel Hill conference was Joseph Weber. Weber was a graduate of the U.S. Naval Academy and a veteran of World War II. As a radar officer, he survived the sinking of the U.S. aircraft carrier *Lexington* in 1942.[61] Weber was then assigned as skipper of a submarine chaser in the Mediterranean Sea, helping to lead the Allied invasion of Sicily in 1943. After the war, Weber studied electronics at the Naval Postgraduate School, and eventually rose to the rank of lieutenant-commander in charge of electronic countermeasures for the Navy.[62] Weber gained expertise in using microwave technology to counter enemy radar. Much of his efforts were on finding better ways to improve microwave signal-to-noise ratios.[63]

Although he had not yet completed a PhD, in 1948 Weber accepted a professorship of electrical engineering at the University of Maryland (UMD). Over the next three years, he taught at UMD while simultaneously completing a PhD in physics at the Catholic University of America. Under the supervision of Keith Laidler, Weber's doctoral research extended his microwave expertise to the study of chemical kinetics in ammonia.[64] But perhaps the most memorable events of his graduate school experience were the lectures on atomic physics given by physicist Karl Herzfeld.[65] In one particularly inspiring seminar, on Einstein's deduction of the Planck radiation law in 1917, Herzfeld discussed stimulated versus spontaneous emission. Weber immediately realized that stimulated emission could be used to make a novel low-noise "microwave amplifier." To his knowledge, no one had ever taken advantage of the quantum states of atoms and molecules to create highly coherent microwave signals.[66]

In addition to his doctoral research on microwave spectroscopy, Weber spent the next several years working in isolation to develop his concept of coherent microwave amplification. In 1951, he submitted a summary of his research to be presented at a 1952 radio engineering conference in Ottawa, Canada.[67] When finally published in the conference proceedings, Weber's summary became the first to describe the concept of microwave amplification by stimulated emission of radiation (MASER)—a forerunner to the now ubiquitous LASER.[68]

Unfortunately for Weber, his paper was not widely circulated. Any attention Weber did receive for his microwave amplification theory was thanks to the Radio Corporation of America (RCA) paying him $50 to present on his research. Charles H. Townes, who had been independently working on stimulated emission, happened to hear his talk at RCA. At the time, it was not clear that Weber's ideas for coherent microwave amplification could be harnessed, but Townes went on to fully develop maser theory and build the first working maser in 1953. Weber seems to have missed the opportunity to join Townes as co-inventor of the maser.[69] The contrast between Townes's Nobel Prize and Weber's $50 paycheck would become a sore point for Weber, and some of his colleagues, who believed he deserved more recognition.[70]

Leaving missed opportunities with the maser behind, Weber searched for other projects.[71] Having earned his first sabbatical in 1955–56, he headed to Princeton, where he had fellowships to work with Oppenheimer and Wheeler.[72] The year was transformational for Weber, shifting his sights to the study of GR and gravitation. Weber found natural resonance with Wheeler, who was one of Herzfeld's most prominent students and who was by then a leading figure in physics.[73] Wheeler had decided to spend the spring semester at the Lorentz Institute for Theoretical Physics in Leiden, and invited Weber to join him.[74] There, Wheeler and his group immersed themselves in the study of GR and its predictions of gravitational waves.[75] The following year, 1957, Weber joined Wheeler at the Chapel Hill conference, becoming part of the small cohort of physicists discussing the possibility of detecting gravitational waves.

During this fertile period—fresh on the heels of the Bondi-Feynman thought experiment—Weber began his own program to detect gravitational waves. After Chapel Hill, Weber and Wheeler wrote a paper in which they too used Pirani's calculations as the foundation for their own thought experiment to prove the existence of gravitational waves. It appeared in the summer of 1957, and was followed by two subsequent papers, in 1960 and 1961, in which Weber took up the challenge of turning these thought experiments into a bona fide research program to detect gravitational waves.[76]

Joseph Weber: Potential Star of Gravitational Wave Theory

For physicists working on GR, Chapel Hill was a watershed. Independently and almost simultaneously, Bondi, Feynman, Wheeler, and Weber converged on thought experiments, built on Pirani's calculations, arguing for the existence of energy-carrying gravitational waves. As the arguments found their way to publication, many theorists in GR shifted their attention to deeper problems. If gravitational waves carry energy, how is this manifested in particular cases, such as orbiting binary systems? Do these systems radiate energy in the form of gravitational waves? And if so, do their orbits decay?[77]

Weber had spent considerable time with Wheeler sharpening his expertise in GR. In Princeton and Leiden, he had been exposed to some of the most ingenious theoretical minds of the time. By the early 1960s, Weber had published a textbook on GR as well as numerous articles; he was considered one of the most knowledgeable physicists in the field.[78] Recall that a decade earlier, Weber had been one of the first to conceptualize the maser, yet failed to capitalize by developing a working prototype. This time he seemed determined to avoid that pitfall. Shortly after he and Wheeler published their version of the Bondi-Feynman thought experiment, Weber began considering how to make these thought experiments reality.

According to Kip Thorne, Weber spent the next two years working on possible experimental designs, filling twelve hundred notebook pages with ideas for gravitational wave detectors.[79] He submitted his evolving ideas to the Gravity Research Foundation and won their 1959 essay contest; his paper was published in *Physical Review* in January 1960.[80] Weber continued to experiment, and, after much trial and error, eventually landed on an optimal detector scheme. By the end of the 1960s, Weber was collecting data from multiple antennas, two of which were 1,400 kg solid aluminum cylindrical bars located more than a thousand kilometers apart.[81] These "Weber bars" were essentially giant tuning forks. The detectors were designed to resonate with incoming gravitational waves of a particular frequency (see Figure 4.8).

In theory, the tidal forces generated by gravitational waves would cause tiny vibrations in the bar, compressing and stretching it as they passed through. According to Weber's calculations, if conditions were just right—if the bar had the correct resonant

Figure 4.8 Joseph Weber working on one of his Weber bar antennas at the University of Maryland, circa 1969. (Source: Courtesy of the University AlbUM Collection, University of Maryland, College Park.)

dimensions and all other sources of internal and ambient vibrations were minimized—the ringing of the bar due to incoming gravitational waves might be detectable. To reduce the noise, the bars were suspended by thin steel wires and enclosed in large vacuum chambers, which were seismically isolated. The bars were partially covered with piezoelectric sensors that could detect and amplify the tiniest strains in the metal (see Figures 4.8 and 4.9).

This was an audacious experiment. Most physicists believed it had no chance of success because the detectors could not be sensitive enough to distinguish a gravitational signal from the bar's internal thermal noise and noise caused by ambient electrical, magnetic, acoustic, and seismic activity.[82] Undeterred, Weber spent the 1960s working with graduate students at his University of Maryland lab to improve his detectors. Many of Weber's innovations would later inspire scientists designing more advanced

Figure 4.9 Weber at work in his lab at the University of Maryland, circa 1969. In the background one can see a large open vacuum chamber he used to isolate his detectors. Inside is one of his disk model antennas. (Source: Courtesy of the University AlbUM Collection, University of Maryland, College Park.)

detectors. For example, his idea to use identical detectors more than 1000 kilometers apart while also introducing artificial time delays into data runs to ensure optimal correlation has been adopted by newer gravitational detection projects like LIGO.[83]

Throughout the 1960s, Weber held the somewhat dubious distinction of being "the only experimental physicist in the world seeking gravitational waves."[84] During this time, he was prolific, publishing more than twenty papers stemming from his research.[85] However, a fine line divides trailblazers from pariahs in science, and Weber certainly navigated that boundary. While most physicists working on GR were skeptical of Weber's detection program, he carried on. Patrons like the NSF and NASA were funding his research; important physicists, including Freeman Dyson and Wheeler, were encouraging, even optimistic, about his chances. In 1967, Wheeler predicted that gravitational waves were going to be one of the "big discoveries of the next ten years." In part due to Weber's work, Wheeler was convinced that someone would soon "detect them for the first time."[86]

Soon enough, Wheeler's prediction seemed prescient. In late December 1968, Weber's detectors in College Park, Maryland, and at the Argonne National Laboratory near Chicago showed a simultaneous flurry of activity. Over the next three months, seventeen more coincident signals would be detected.[87] Throughout the winter and spring, Weber and his team "went through an exhaustive process to make sure the signals were real." They inserted time delays, checked their equipment in every conceivable way, looked for possible electromagnetic interference effects—including solar flares and lightning strikes—monitored seismic activities and cosmic ray effects. Nothing seemed to explain the coincident signals at the two detectors.[88] Although the signals were far greater than he, or anyone else, expected, Weber concluded that they must be due to gravitational waves.[89]

At a GR conference held in Cincinnati in June 1969, Weber could contain himself no longer. According to Kip Thorne, Weber shocked his colleagues by standing to announce the detection of gravitational waves. At first, people took these claims very seriously. Weber's announcement was even greeted with "applause and tributes." Two weeks later, his peer-reviewed publication appeared in *Physical Review Letters.* Weber's detection of gravitational waves had become a sensation, seemingly one of the most important discoveries of the twentieth century.[90]

Becoming a Gravitational Wave Pariah

According to Harry Collins's decades-long sociological study, the operatic melodrama that followed Weber's claims destroyed both "lives and spirits."[91] From Collins's thoughtful analysis, one senses that theory and experiment alone were insufficient to settle the question of whether Weber had indeed detected gravitational waves in 1969.[92] Eventually, by consensus, the scientific community answered this question with a definitive "No." However, Collins's examinations call into question exactly how that consensus was reached. To understand the fate of Weber and his claims of detection, one must understand what Collins calls the ripples in "social spacetime" that followed the claims.[93]

We saw in Chapter 1 that Collins's sociological study of consensus differs in key points from that of physicist Allan Franklin's.[94] I wholeheartedly agree with their groundbreaking collaborative essay exploring their decades-long debate.[95] I tend to read the two analyses as complementary. Beyond their differences on epistemological assumptions, the real difference between Collins and Franklin is in methodology. Collins relies heavily on fieldwork, including interviews over several decades with many of the participants. Franklin's analysis is more internalist, focusing on the published literature. Collins's 2004 account was published after Franklin's first account and responds to many of Franklin's earlier critiques. Each approach paints a different picture of the controversy; reading both yields a more holistic picture of events.

As a field, gravitational wave detection suffered through an inflationary period after Weber's shocking 1969 announcement.

When Weber's published results showed signals that were orders of magnitude stronger than predicted, some colleagues were eager to verify his results.[96] Suddenly, teams from all across the world began setting up to test Weber's claims. By 1972, twelve or so groups were checking Weber's findings.[97] These early attempts, however, failed to show positive results, and were generally considered inconclusive. In a review article on gravitational-wave astronomy in 1972, William Press and Kip Thorne were ambivalent in judging Weber's results. Although they found Weber's evidence for gravitational waves "fairly convincing," it was, they noted, inconsistent with currently accepted theoretical models.[98]

Considering the extremely low signal-to-noise ratio, gravitational wave detection experiments are notoriously complicated and difficult to execute. As a result, uncertainty dominated the field for years. Collins identifies the difficulty of other scientists' directly testing Weber's detection as "experimenter's regress."[99] This conundrum is best captured by the question: How does a scientist know whether an experiment has worked? Someone replicating a known experiment with predictable, understood results can compare their data to the established norm of expected results. That comparison will immediately tell whether their experiment succeeded. But for truly novel experiments without established norms and expectations, how does an experimenter know whether they have succeeded in detecting an effect? One might try to use theory as the litmus test. But what if—as in the case of gravitational waves in the early 1970s—the theory, too, is somewhat uncertain? How could Weber, or anyone else, be sure they had detected gravitational waves?[100]

At the time, the field of gravitational wave detection was still nascent. Weber's announced detections had enticed scientists from different specialties to visit this fledgling subdiscipline of physics, bringing with them a multitude of frameworks and different forms of expertise. There were no universally agreed-upon procedures and standards to judge by. But, as the earliest attempts to experimentally confirm Weber's results failed, the tide began to turn from attempts to confirm to attempts to falsify. Most physicists did not attempt to replicate Weber's experiments. Instead, they set up experiments that could probe his results indirectly, each in their own way.

It was a messy affair. The skeptics who challenged Weber's results in those early years debated the validity not only of his results but of one another's analyses.[101] Overall, their work lacked coherence. According to Collins, the "output of [their] experiments was not as uniform as might be thought when we look back on them with hindsight, nor is the criticism against Weber so uniformly damning as it now seems."[102] Yet by 1975, the scientific community's rhetoric about Weber and his claims had noticeably shifted. By then enough articles had been published claiming to falsify Weber's results that the scientific consensus began to congeal; the verdict had all but evolved that Weber had not detected gravitational waves. Weber, meanwhile, was frustrated by the lack of direct replication of his experiments. He continued to respond to critics by refining both his detectors and data analysis.[103]

The case against Weber was prosecuted by a small number of scientists, most aggressively by Richard Garwin, a fellow at IBM's Thomas J. Watson Research Center. Garwin took particular issue with Weber's failure to publicly admit clear errors in data analysis.[104] In September 1973, Weber and others published a paper in *Physical Review Letters* based on their observations of gravitational wave detections. Included were analyses of coincident counts thought to be observed simultaneously at detectors at the University of Maryland and the University of Rochester. In subsequent conversations with collaborators in Rochester, Weber realized that among other computational errors, his claim of simultaneous detection at the two distant labs was problematic. When he originally analyzed the results, Weber had not known that the two labs used different time zones to calibrate their data. As a result, Weber had failed to adequately synchronize the data runs, and was conjuring signals out of pure noise. This was an understandable error, the sort that happens in experimental

physics. What really angered Garwin and others, though, was
Weber's refusal to publish a note publicly admitting to this and
other errors in the published paper.[105]

As the calendar turned to 1974, Weber preferred to focus on
new detection results rather than harp on old errors in one particular data run. He believed that he had addressed his critics' concerns with improvements to his experiment. Instead of publishing
a retraction, he pushed forward, continuing his gravitational wave
detection research. Garwin would not let it go. He began a relentless campaign to discredit Weber's work. The dispute escalated
from a technical dispute in a subfield of physics to a full-blown
controversy, involving many in the broader physics community.
In December 1974, Garwin published a bluntly hostile exposé of
Weber's errors in the broadly circulated *Physics Today*. Throughout
1974–75, Garwin used publications, conference talks, and extensive
private correspondence to doggedly pursue Weber.[106]

Ultimately, Garwin saw Weber's gravitational wave experiments
as an example of what Irving Langmuir called "pathological science."[107] By emphasizing this label, highlighting the errors in
analysis in the 1973 *Physics Review Letters* paper, and pointing out
Weber's stubborn refusal to acknowledge and retract errors,
Garwin felt he could discredit all Weber's previous detection
claims. Doing so might save the scientific community from a protracted, messy controversy. From Weber's perspective, Garwin
had challenged the validity of one particular experimental run,
but couldn't invalidate all of his findings. Weber was continuously
improving his experiments and analyses; he expected to conduct
a prolonged, open disciplinary discourse to determine the validity of his larger research program.

According to Collin's analysis, the problem of experimenter's
regress appears to have been solved in the Weber case without
undisputable evidence. No single experiment or theoretical calculation falsified all of Weber's detections. The rejection of his
work leaned heavily on *social* arguments such as scientific reputation, interpersonal dynamics, and ethics of publication and

retraction. As multiple colleagues of Garwin admitted to Collins in later interviews, Garwin's aggressive campaign to discredit Weber departed significantly from traditional forms of scientific research. His campaign had migrated to the realm of social space-time.[108] Nevertheless, persuaded by Garwin's attacks and other analyses showing negative results, many within the scientific community began to accept the notion that Weber's work was pathological and should be ignored.

By the end of 1975, "nearly all scientists agreed that Weber's experiment was not adequate," even though their reasons for believing this "differed markedly."[109] Some saw errors in the computer program designed to identify signals; some questioned the statistical analyses necessary to distinguish signals from noise; still others found fault in the simultaneity of the two detectors' coincident counts. Whatever their reason for rejecting Weber's claims, most scientists saw his results as highly problematic. As a result, his once-promising approach to gravitational wave detection became unwelcome. The post-Weber era of gravitational wave detection had begun.[110] It was marked not by a rejection of Weber's ideas, but by their complete omission. Except for brief episodes in the late 1980s and early 1990s, Weber and his work were mostly ignored, purged from mainstream scientific discourse. He had become a scientific pariah, on the verge of being forgotten.[111]

Phoenix Rising: Rehabilitated Scientific Hero

Feeling wronged by Garwin and others in the scientific community, Weber spent a quarter century bitterly and staunchly defending his gravitational wave results.[112] Although his sources of funding were diminished, he continued to claim to be using ever more sensitive detection equipment and rigorous analytical methods. Even after the NSF completely cut off his funding in 1987, he persisted, finding new allies and patrons.[113] When the broader community began to use more elaborate equipment—including cryogenically cooled resonant bars and large-scale interferometers—Weber defiantly clung to his approach based on room-temperature resonant bars.[114]

As the scale of LIGO's interferometry reached "big science" levels of funding, Weber lashed out. Why should the NSF devote hundreds of millions of dollars to develop an uncertain technology, when Weber had for decades been successfully detecting gravitational waves using small-scale, room-temperature resonant bars? Wouldn't tax dollars be better spent investing in proven technology, at a fraction of the cost? For Weber these questions weren't merely rhetorical. In the early 1990s, he began lobbying Congress to stop funding LIGO, attacking both the premise and the leadership of the LIGO program.[115]

Weber's critiques were not just self-serving exercises to draw attention to himself. He was fighting an existential scientific battle. In a post–Cold War environment of restricted funding for science, LIGO was squeezing out small-scale projects like his. For decades, Weber had weathered the skepticism of his peers thanks to continuing, albeit dwindling funding streams. The NSF's decision to fully fund LIGO meant that Weber's smaller program would be starved to death.[116] From Weber's perspective, he was being forced out of gravitational wave detection research for political expediency, without a fair hearing.

Weber's tenacity raised the ire of some LIGO scientists, who thought he was sowing enough doubt to jeopardize their program. Although they had ignored him for nearly two decades, they began to briefly re-engage Weber's work with the sole purpose of undermining his credibility and pushing him to the margins for good.[117] Weber never recovered. He died in 2000 without having convinced most of his colleagues of the validity of his gravitational wave detection claims.

Yet something funny happened on the way to scientific oblivion. During the first two decades of the new millennium, Weber's legacy as a pioneer of gravitational wave detection was completely rehabilitated. The prominence he failed to achieve during his lifetime finally came his way. This transformation was not the result of a happy historical accident. It was a case of strategic myth-historical reconstruction, coming from an unlikely source. The same LIGO team that Weber had critiqued for years was now determined to reimagine his role in the narrative of their triumph!

On February 11, 2016, at the National Press Club in Washington, DC, in a roomful of reporters and dignitaries, Kip Thorne declared Joseph Weber's work on gravitational wave detection "pioneering." NSF Director France Córdova repeated the term "pioneering," then mentioned Weber in the same breath as Nobel laureates. At one stroke, Weber—the social pariah who had been ridiculed for decades—was transformed into the physics' latest hero; the Beast became the prince. Without changing any of his actions, scientists associated with LIGO suddenly embraced Weber as a visionary, brave, and persistent pioneer.

Weber's past was quietly rewritten. The controversy over his claimed detections was erased, and Weber was recast as a father of gravitational wave astronomy. Córdova announced that one of Weber's early detectors was on display at the LIGO site in Hanford, Washington. It was as if an altar had been erected around a relic made sacred by an emerging myth-history. It was a celebration of a scientific hero who had died sixteen years earlier, a cautionary tale against pathological science. Weber's widow, Virginia Trimble, was an honored guest of the LIGO team at their celebratory news conference in February 2016.

Trimble, herself an astronomer and historian of science, has been adamant that the creation of LIGO should not have required eliminating other research programs like Weber's: "I think if there had been two technologies going forward they would have pushed each other, as collaborators not as competitors....It might have led to an observation sooner."[118] After his death, Trimble says she made donations in Weber's name out of "anger," presumably over his treatment by the scientific community.[119] For example, when she sold their home in Maryland, she used part of the proceeds to endow the American Astronomical Society's annual Weber Award for Astronomical Instrumentation.[120]

One wonders if Trimble's social capital, including her long friendships with scientific leaders like Córdova, helped the rehabilitation process along.[121]

To his credit, Thorne seems to have maintained a consistent respect for Weber. In 1977, when Weber was becoming a scientific pariah in the GR community, Thorne contributed a chapter to a volume Weber co-edited.[122] Then, almost two decades later, amid Weber's campaign to derail the LIGO program, Thorne spoke fondly of him in his bestselling book *Black Holes and Time Warps*.[123] Thorne and many of his colleagues seemed genuinely to want Weber recognized as a pioneer. But they had to bide their time, waiting for him to stop interfering in the establishment of his own legacy. As one physicist said, if he had just let things go and admitted his mistakes, he might have become old "Uncle Joe," the "venerable leader of [his] field."[124] Unfortunately, death seems to have been the necessary precondition for Weber's legacy to be acknowledged.

Whatever the reasons for this particular myth-historical reconstruction, Weber is now a certified hero. Knowing about the reconstruction, one might question whether Weber's hero status is justified. There is no doubt that, historically, he was one of the most important pioneers in gravitational wave detection. The heat generated by the controversy surrounding his claims of detection had surprising, unintended consequences. Although Garwin left the field of gravitational wave detection, others did not. The focus shifted from falsifying Weber's detection results to developing ever more sensitive gravitational wave detection schemes. LIGO was the indirect result of the mid-1970s detection controversy. Ironically, it was Weber's own graduate student who developed one of the earliest interferometers, resurrecting a scheme Weber had considered, and abandoned, in the early 1960s.[125]

All these caveats aside, the myth-history spun by Thorne, Córdova, and the LIGO team remains problematic. There is no doubt that Weber should get his historical due and not be forgotten. However, in elevating Weber, the LIGO team has filtered out the controversy surrounding his gravitational wave detection results. They have omitted all social dynamics that fail to represent ideal norms of scientific practice and discourse. As Franklin notes, the case of Weber "demonstrates that 'universalism' as both a norm and an ideal of science is rather complex in its application."[126] This case of myth-historical reconstruction both simplifies complex social dynamics and purges inconvenient facts from the record. The scientist-storytellers at LIGO have created a myth-historical narrative that is more caricature than reality.

One may ask, What harm does such distortion cause? After all, the narrative redresses an injustice in the history of science, and gives students and the public an ideal to aim for when conducting or assessing science. The problem is that as long as people learn the history of science through pristine myth-histories that replay idealized tropes and celebrate impossibly perfect heroes, science will remain unidentifiable and inaccessible to most audiences. This in turn makes the projects of humanizing and diversifying science more difficult. If we are to fix the "pipeline" problem by building a more diverse and inclusive scientific community, relying on stale myth-histories will not help. Standpoint epistemologists argue that a strong consensus is best forged by a community that is diverse in its constituents. It follows that rethinking the stories we tell and how we curate them is critical.

Another victim of myth-histories is people's trust in science. When the public is constantly told idealized myth-histories, then confronts a realistic picture of messy science in the making, there is confusion and cognitive dissonance. How to make sense of this mismatch? How can people trust science when the portrayal of science is false? The following chapter explores a life-and-death example of why public trust matters.

5

Demarcating Seismic Uncertainties

A Front in the "War on Science"

> Everything faded into mist. The past was erased, the
> erasure was forgotten, the lie became truth.[1]
> —GEORGE ORWELL

Grappling with Uncertainty

Uncertainty—an inevitable feature of the human condition—
can cause stress, even overwhelm at times (see Figure 5.1).
Humans have created cosmologies to help us deal with uncertainty by producing knowledge that allows us to draw the cur-
tains on unsettling darkness. Historically, cosmologies have
taken many forms: mythologies of gods and heroes; religious
concepts of divinity; rational universal ontologies; modern scientific theories. Although these frameworks are not necessarily
mutually exclusive, for many scientists and their allies, modern
science has become the only true candle in a sea of darkness.[2]

This view of scientific exceptionalism too often spurs aggressively protectionist campaigns in response to perceived threats to
scientific authority—indeed, anything that challenges the ideal
image of science. Leon Lederman's attack on pseudoscience in his
book *The God Particle,* discussed in Chapter 1, showed how protectionist campaigns tend to overcompensate, drawing demarcations
too clearly between scientific knowledge and all other forms of
knowledge. This boundary work is a surefire sign that scientist-
storytellers are using myth-historical rhetoric to filter out noise,

Figure 5.1 Paralysis of uncertainty can be real. Scientists are constantly grappling with uncertainty. How can they know if the results of their research are enough to determine a particular conclusion? They are often engaged in varying levels of "underdetermination" in which they struggle to eliminate alternative explanations, with the goal of matching evidence with a determined conclusion. (Source: Illustration by Zeyu [Margaret] Liu.)

and elevating a particular image of idealized science. Once recast by rhetorical filtration and selection, a constellation of facts can become fungible, bending to rhetorical priorities.

Theories that have not always been fully vetted by the scientific community, but are rhetorically useful, are sometimes prematurely certified as consensual, then used to discredit unwanted ideas. Defeated ideas are then dismissed, discarded to "the debris of history's footnotes."[3] When they pose an imminent threat to scientific authority, they can even be prematurely declared "impossible."[4] Such rhetorical pruning of threatening ideas can extend to their advocates as well. Rhetorical campaigns are routinely mounted to undercut the scientific legitimacy of adversaries; sometimes, at their strongest, acts of demarcation become exorcisms.

The demarcation problem is intractable. There is no established list of criteria, no universal method, no prescriptive norms of

practice that unambiguously delineate science from pseudoscience. As we have repeatedly seen in this book, any careful analysis of scientific practice will inevitably reveal human and social
elements that make it appear messy and incongruent with science's myth-historical ideals. Nevertheless, noisy details that
myth-historians filter out of their storytelling are critical to a
fuller understanding of science.

This chapter explores the tension inherent in the juxtaposition of scientific ideals and concrete scientific practice as it relates
to managing scientific uncertainty. Scientists seem to understand
their work within the larger canopy of uncertain knowledge,
though they may not always present it that way. They know that
their data and their analyses should be understood within a relative, probabilistic context of graded uncertainty. How do scientists know their data represents a coherent signal? How do they
avoid unconscious bias toward overfitting their data, mistaking
noise for signal?

Conversations among scientists about the veracity of their
conclusions tend to include quantifying their confidence levels
with calculations of probability (p) values and their corresponding variability (sigma) values. However, recent controversies
suggest that scientists often misuse statistical analyses, and nonscientists often misunderstand measures of uncertainty.[5] One of
the most telling controversies calls into question the reliability of
scientific reproducibility, a foundational pillar of science. The
issue is so critical that a group of scientists felt compelled to write
a "Manifesto for Reproducible Science."[6]

That lay people misunderstand scientific uncertainty should
not surprise us, given that most people's exposure to science is
limited to introductory classes and myth-histories. Here, the science tends to be oversimplified and the concepts are straightforward, right or wrong. The answers are already known: students
are led confidently to the correct answer. Similarly, teaching labs
are performative: they demonstrate known facts rather than
genuinely explore the unknown.

Of course, in reality, scientific practice is much messier than what's portrayed in such settings. Scientists work in relative darkness, struggling constantly with the unknown, and tackling complex problems without clear, unambiguous answers.

More surprising is that scientists themselves at times misuse statistical analysis and misrepresent uncertainty. A May 2016 article in *Nature* surveyed more than fifteen hundred scientists from many disciplines on their thoughts about scientific reproducibility. Of those, over 70% had "tried and failed to reproduce another scientist's experiments," and 90% thought that science was mired in a "reproducibility crisis" (see Figure 5.2).[7] The three most commonly cited factors that contributed to the reproducibility failures were selection bias, pressure to publish, and ineffective statistical analysis.[8] So scientists admittedly fall into cognitive

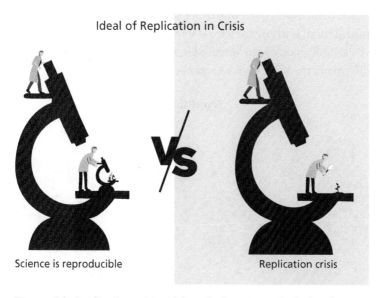

Figure 5.2 Replication crisis. Although there is an ideal of replication in science, there has been a recent "crisis" in some domains in which many results have been shown to be not exactly reproducible. (Source: Illustration by Zeyu [Margaret] Liu.)

traps that produce false-positive signals out of uncorrelated noise, then justify these by misusing statistical analyses.[9]

As some statisticians have noted, "A core problem is that both scientists and the public confound statistics with reality. But statistical inference is a thought experiment, describing the predictive performance of models about reality."[10] Statistical results can be misleading and problematic if they are treated as representations of reality rather than a simplified model of it. When scientists perform experiments and collect data, they must distinguish significant signals from irrelevant noise. Statistical analysis is commonly used to do this, but overreliance on statistical manipulation without reflective analysis of underlying assumptions can yield distorted claims of statistical significance and a false sense of certainty.[11]

In *The Signal and the Noise*, Nate Silver summarizes the conundrum that stems from scientists sifting through too much data:

The instinctual shortcut that we take when we have 'too much information' is to engage with it selectively, picking out the parts we like and ignoring the remainder, making allies with those who have made the same choices and enemies of the rest.[12]

Overfitting

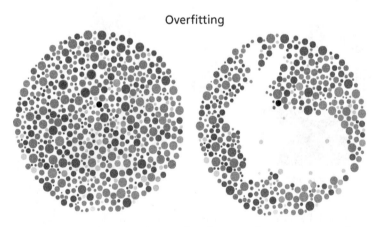

Figure 5.3 Overfitting. In statistical analysis, overfitting can occur when a signal is mistakenly seen in a pattern of pure noise. Here we see a rabbit emerge from the noise. (Source: Illustration by Zeyu [Margaret] Liu.)

Thus, selection bias can cause scientific discourse to provide misleadingly clear-cut answers. On the other hand, Silver notes that when data is extremely limited, as in observational astronomy and seismology, cognitive bias can lead to overfitting (see Figure 5.3). In this case, scientists can read too much into their data, overinterpreting signals out of noise.[13] These conundrums seem at the heart of the reproducibility crisis, foregrounding the human and social dimensions of science as practiced.

War on Science

From discussions of the reproducibility crisis, we see that managing and communicating uncertainty is a fundamental problem in science. It is one that is not considered rigorously enough. Scientists seem confident they are bridging Einstein's characterizations of science: taking a subjective, psychologically conditioned human practice and working to extract from it objective, certain scientific knowledge.[14] But this representation is rooted in a tension that remains latent and unresolved. It appears most clearly during scientific controversies, when the social and human dimensions of science become all but impossible to ignore.

In order to dismiss recent controversies in science, some observers have tied these into a larger social movement that threatens to shred the very fabric of science. It appears that a wide-ranging, irrational war is being waged on science. Fronts in this war appear on national and international stages and on all sides of the political landscape, where the emergence of a "post-fact" politics "has normalized the denial of scientific evidence" when it conflicts with personal agendas.[15] Whether someone rejects the incontrovertible evidence for anthropogenic climate change or dismisses studies showing no causal links between childhood vaccinations and autism, some segments of the population regularly dismiss consensus-driven scientific evidence and authority.

It's not clear that a formal, coordinated war on science is underway, but there does appear to be a significant erosion of confidence in scientific institutions and their authority. While no single cause can be identified, most of the surrounding discourse fails to address how science and its practitioners are complicit. Instead, the analyses place most of the blame on postmodern academics, for promoting extreme forms of relativism; on journalists, for giving equal footing to all sides of debates; on industrialists, for exploiting science and technology to increase their personal and corporate wealth; and on politicians, for wielding science as a weapon in order to win power and drive social change.[16]

I do not doubt the influence of these various factors. That said, we should also ask how scientist-storytellers themselves contribute to a fundamental misunderstanding of science. This chapter looks closely at a scientific controversy in which the notion of uncertainty played a leading role. Based on our exploration of the power of scientific myth-histories, we examine the ramifications of a case in which scientists tried to manage public anxieties over uncertainty, while safeguarding and promoting their myth-historical ideals.

Before the L'Aquila earthquake in spring 2009 (see Figure 5.4), a group of Italian policymakers and scientists engaged in aggressive acts of public demarcation between science and pseudoscience. In the wake of the ensuing destruction, survivors, bolstered by Italian courts, blamed the deaths of their loved ones on these same scientists and policymakers. Scientific failure did not arise from the particulars of the seismic models used. Failure occurred at the intersection of the public's misunderstanding of science and the scientific community's defense of myth-historical imaginaries. The fault lines in this case run much deeper than public officials' trying to manage public anxiety about seismic uncertainty. As one observer noted, "The modern ideal of scientists (alone) 'speaking truth to power'... ended up collapsing together with the buildings of L'Aquila."[17]

Figure 5.4 The destruction to city of L'Aquila caused by the 2009 earthquake. More than 300 people died and thousands of homes and buildings were destroyed. President Barack Obama, joined by Italian Prime Minister Silvio Berlusconi, tour earthquake damage, July 8, 2009. (Source: Official White House photo by Chuck Kennedy; Official White House Photostream; P070809CK-0208, public domain https://commons .wikimedia.org/w/index.php?curid=8094739.)

Uncertain Shockwaves

In the early morning hours of April 6, 2009, a magnitude 6.3 earthquake shook the Abruzzo region in central Italy.[18] The earthquake, its epicenter 3.4 kilometers southwest of the city of L'Aquila, left more than three hundred people dead, more than fifteen hundred injured, thousands of homes and historic buildings destroyed, and the communities reeling.[19] Families were torn apart and the region found itself digging out from damages that reached tens of billions of euros.[20] In ancient times, such a tragedy might have been blamed on angry gods, but in October 2012, it was seven scientists and public officials who were found legally responsible for dozens of deaths.[21]

Judge Marco Billi explained his ruling by noting that the "L'Aquila Seven," working as part of an official risk assessment advisory committee, were guilty of practicing and communicating science that was "approximate, generic, and ineffective in relation to the activities and duties of forecasting and prevention."[22] The committee members were convicted of multiple counts of manslaughter, each sentenced to six years in prison, and ordered to pay more than $10 million in damages to earthquake victims.[23]

At first glance, this seems a grave injustice, and much initial reaction from scientific communities around the world was disbelief and outrage. The CEO of the American Association for the Advancement of Science (AAAS), Alan Leshner, captured much of the international outrage in a letter of direct appeal to the president of Italy. Leshner implored the Italian president to intervene, claiming that the charges brought against the six Italian scientists were "both unfair and naïve."[24] According to Leshner, the indictments were based on unrealistic expectations of scientific seismic prediction. The scientists were accused of "fail[ing] to alert the population of L'Aquila of an impending earthquake." But, as Leshner pointed out, "There is no scientifically accepted method for earthquake prediction," so the scientists could not possibly have warned citizens of an earthquake with any scientific credibility.[25]

Leshner told the president that the scientific community was deeply concerned that

> subjecting scientists to criminal charges for adhering to accepted scientific practices could have a chilling effect on researchers, thereby impeding the free exchange of ideas necessary for progress in science and discouraging them from participating in matters of great public importance.[26]

Leshner posed a straightforward question: how could the science of seismology be blamed for the death and destruction in L'Aquila when, according to scientific consensus, earthquake prediction was currently impossible? To many observers, the trial was symptomatic of a broader assault on science. Convicting the scientists seemed a regression to medieval law—the twenty-first century equivalent of Galileo's trial at the hands of the Inquisition in 1633.[27]

Readers who bypass the sensational headlines comparing the L'Aquila convictions to the Inquisition soon realize that Judge Billi's rulings did not emerge from an arbitrary or reactionary medieval adjudication. There is much more to this story than Leshner's letter suggests. From the perspective of many L'Aquila residents who had their world upended on April 6, the story centers on unavoidable feelings of "betrayal" by science.[28] What seems clear to those involved in the court case is that the group was not held accountable for failing to *predict* the earthquake. Instead, their legal liability rested on a failure to carry out their official responsibilities. As appointed public servants with clearly articulated tasks, they failed to assess and communicate seismic risks, and to inform the public of best practices to prevent injury and death.

If the L'Aquila Seven weren't tending to their official responsibilities, what were they doing? In the lead-up to the earthquake, public officials were focused on an aggressive, coordinated media campaign. The campaign had two apparent goals: (1) to discredit a local man who claimed he could scientifically predict earthquakes; and (2) to reassure local residents (*Aquilani*) that ongoing seismic activity was normal and there was nothing to fear. However, in their zeal to achieve these goals, the Italian National Department of Civil Protection (DPC) failed to accurately represent scientific consensus and painted a distorted picture of scientific uncertainty.[29] The authorities protected scientific boundaries instead of their fellow citizens.

Uncertain Knowledge(s)

Early press coverage of the L'Aquila Seven convictions framed the controversy as Leshner's AAAS protest letter had, claiming a rogue Italian judicial system unfairly put science on trial. If seismic prediction was impossible, how could scientists be indicted for failing to predict the L'Aquila earthquake? This framing filters out important details of what happened in L'Aquila in the months before the earthquake.

Vincenzo Vittorini, a L'Aquila surgeon, lost both his wife and nine-year-old daughter in the earthquake. He felt "betrayed by science."[30] He found the scientists' and officials' statements before the earthquake to be "dramatically superficial," distorted, and falsely reassuring. According to Vittorini, the *Aquilani* were "deprived" of their "fear of earthquakes," thereby betraying "the culture of prudence and good sense that our parents taught us on the basis of experience and of the wisdom of the previous generations."[31]

From Vittorini's perspective, many deaths were preventable—not because the scientists could control or predict the imminent earthquake, but because their misleading public proclamations created an "anesthetizing," false sense of security.[32] Many L'Aquila residents abandoned their rich heritage of traditional knowledge, the wisdom of generations of families living in a seismically active region. Through a long history of seismic activity, L'Aquila had traditionally prepared local residents to be vigilant, to respond to nighttime tremors by sounding alarm bells and "sleeping in cars or staying up in the piazza."[33]

However, when a strong foreshock struck L'Aquila just before 11 p.m. on April 5, 2009, Vittorini, with many other residents, found himself anesthetized by the persistent, confident scientific proclamations minimizing imminent risk. The surgeon ignored the local culture of vigilance and any instincts he had to spend the night outdoors. When his wife, Claudia, fearful of the ongoing tremors, asked if they should spend the night outside, Vittorini responded: "But Claudia, at this point the release of energy has happened! It's like the experts said, there won't be stronger shocks, so we can stay calm!"[34] Based largely on expert scientific advice, Vittorini convinced himself and his wife to stay in their apartment. As the night wore on, they kept an eye on the INGV webpage and their television tuned to TVUNO to monitor the latest news.[35]

When the main shock struck less than five hours later, Vittorini was in bed, huddled with Claudia and their nine-year-old daughter, Fabrizia. The ground shook violently for "twenty-two devastating seconds."[36] Like thousands of other buildings in L'Aquila, Vittorini's apartment building collapsed in a heap of dust and rubble (see Figure 5.5). His third-floor apartment was crushed under tons of reinforced concrete. Six hours later, Vittorini was pulled from the wreckage alive; Claudia and Fabrizia were dead.[37]

Although the convictions of the L'Aquila Seven were eventually overturned,[38] it is still instructive for our analysis to look closely at the context and actions that led to the original court case. Scientists and public officials involved in the case engaged in more than Leshner's myth-historical ideal of "adhering to accepted scientific practices" and "a free exchange of ideas necessary for progress."[39] These descriptions in Leshner's letter to the Italian president echo myth-historical

Figure 5.5 Structural damage to buildings caused by the 2009 earthquake was extensive. (Source: Photograph by Raffaele Scala; own work, CC BY-SA 4.0. https://commons.wikimedia.org/w/index.php?curid= 48114047.)

ideals of scientific practice that all essentialist representations put forward.

The boundary work performed by scientists and public officials before the L'Aquila disaster consisted of four distinct activities. First, by filtering out the history of legitimate scientific inquiries into seismic prediction, they stigmatized such activity as illegitimate. Second, they explicitly distinguished between *legitimate* seismic forecasting research performed by officially appointed scientific and technical experts, and *illegitimate* predictions by a rogue pseudoscientist. Third, they mischaracterized scientific consensus about seismic risk in attempting to reassure the *Aquilani*. Fourth, boundary work was carried out afterwards to assign varying levels of blame to the L'Aquila Seven.

No matter how much scientists try to control the image of science by purging it of human dynamics, public controversies like L'Aquila restore attention to them. L'Aquila was an attempt to demarcate science from pseudoscience in hopes of diffusing public anxiety about living with uncertainty. The case shows how rhetoric can both distort scientific reality and undermine public trust. The goal of my analysis is not to readjudicate the case, but rather to understand the impact of boundary work that employs myth-historical filtering to control a public narrative.

As noted, the failure in this case was not that the exact time and location of the earthquake was not predicted. Instead, the failure was the scientists' (and other officials') failure to communicate risk. Because they were preoccupied with undermining a purported pseudoscientist, they failed to warn the public of the genuine seismic risk they faced.

A Telling History of Seismic Prediction

The difference between *prediction* and *forecasting* in seismology is critical to what caused this scientific controversy in the first place. The difference between these two forms of prospective statement has a long history.[40] Prediction may be defined as a statement that is deterministic with respect to the time, place, and magnitude of future seismic activity. Forecasting, on the other hand, is associated with probabilistic statements.[41] In recent years, many seismologists have distinguished between the impossibility of earthquake prediction and the practical uses of seismic forecasting, but the distinction has not always been observed.[42]

How to predict earthquakes has been on people's minds since they first experienced shifting ground beneath their feet. A pioneer of modern seismology, Robert Mallet, addressed this in 1873, asking, "Can the moment of the occurrence or the degree of intensity of earthquake shock be predicted?"[43] He made clear that this was impossible in his time, but foresaw the possibility of seismic prediction within a "century or two." With Alfred Wegener's 1912 continental drift theory, scientists began to seriously debate and eventually zero in on the geophysical mechanisms that could explain earthquakes. However, not until the mid-1960s did consensus settle on Wegener's original ideas and a "synthetic, quantitative" theory of plate tectonics.[44]

It is important to note that the geophysical revolution occurred in the context of the Cold War. To help monitor the 1963 Nuclear Test Ban Treaty, seismologists set up global networks of detectors that continuously analyzed earthquake activity. In addition, powerful new computers were used to determine "epicenters, focal depths, and origin times for earthquakes."[45] With robust funding streams and newfound windows into the geophysical mechanisms of earthquakes, the science of seismology flourished. Far from mythological actions of angry gods, seismologists now understood earthquakes as the result of tectonic plates rubbing, interlocking, and sliding past each other. They began to unravel the complex web of underlying factors that characterize the release of seismic waves.[46] The scientific community's confidence in fulfilling Mallet's predicted timeline grew.

The second half of the twentieth century saw a relentless campaign to pursue earthquake prediction based on new understandings of geophysical mechanisms, statistical analyses of historical seismic data, and continuous monitoring of a growing list of possible diagnostic precursors. The science of seismology seemed determined to arrive at the Holy Grail of predictive models.[47] Seismologist Susan Hough describes a series of "pendulum swings" between optimism and pessimism over the possibilities of earthquake prediction.[48] The high point may have come with the apparent prediction of the 1975 Haicheng earthquake in China, which saved "untold thousands of lives." It was later revealed that these predictions were overstated and based in part on abnormal snake behavior, but, at the time, the fear of an "earthquake gap" drove well-funded research to develop predictive models.[49]

By the late 1990s, the pendulum had swung the other way, as one after another seismic prediction model was proposed and failed. Frustrated at the lack of progress, mainstream seismologists stopped using the term "prediction," and, according to Hough, left a void quickly filled by cranks and pseudoscientists.[50] By 2009—the year of the L'Aquila earthquake—mainstream seismologists in Italy unequivocally insisted that seismic prediction was impossible.

Any hint of the pendulum swings in the history of earthquake prediction was filtered out of public discourse in the lead-up to the L'Aquila earthquake. One might argue that myth-historical filtering was necessary in order for public officials to maintain public safety. After all, twists and turns from the scientific past should make no difference to the advice given by experts, which presumably takes into account the most up-to-date knowledge. However, L'Aquila shows that this type of rhetorical omission can have serious unintended consequences. Attempts to manage public anxieties and silence pseudoscientific predictions can—and did—cascade into public mistrust, resentment, and controversy.

Boundary Work: Stigmatized Prediction versus Legitimate Forecasting

Seismic predictions are generally associated with public declarations of alarm that are eventually proven true or false, depending on whether the activity happens as predicted. Seismic forecasts, by contrast, acknowledge the complex, probabilistic nature of the underlying phenomena that cause earthquakes; they are far more cautious about their limited predictive power and accuracy. At the time of this writing (2021), the U.S. Geological Survey answers the question "Can you predict earthquakes?" with a flat "No."[51] Seismic forecasts are akin to weather forecasts: scientific models of complex underlying conditions permit only probabilistic statements about what might happen in the short term.[52] Contemporary scientists who publicly address the difference between prediction and forecasting are effectively doing "boundary work": they are trying to draw a line between scientific and pseudoscientific practices.

The USGS website states reasons that predictive statements about earthquakes are necessarily false. Predictions are not based on "scientific evidence." "Earthquakes," they quip, "have nothing to do with clouds, bodily aches and pains, or slugs." People who try to predict earthquakes are pseudoscientific cranks who rely on unscientific signs; scientists who try to forecast earthquakes are legitimate authorities who use meticulously collected evidence (see Figure 5.6). The USGS statement continues:

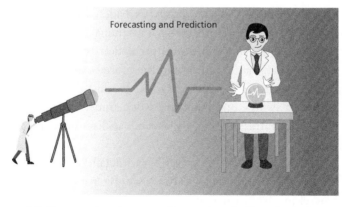

Figure 5.6 Demarcation between forecasting (left) and prediction (right). Although having a rich, and recent, history of sanctioned predictions, the seismology community has decided to make a clear demarcation, stigmatizing prediction. (Source: Illustration by Zeyu [Margaret] Liu.)

Predictions (by nonscientists) usually start swirling around social media when something happens that is thought to be a precursor to an earthquake in the near future. The so-called precursor is often a swarm of small earthquakes, increasing amounts of radon in local water, unusual behavior of animals, increasing size of magnitudes in moderate size events, or a moderate-magnitude event rare enough to suggest that it may be a foreshock.[53]

The statement is fascinating, for if you remove the references to "nonscientists" and "social media," in the mid-1990s this comment could have been found in a scientific journal, grant proposal, or symposium sponsored by the U.S. National Academy of Sciences, when many mainstream scientists were still working to find diagnostic precursors of earthquakes that might well lead to predictive models.[54]

With each pendulum swing between optimism and pessimism, it becomes more difficult to demarcate legitimate scientific pursuits from pseudoscience. The USGS's myth-historical claim of a clear demarcation does not appear to be sustainable. In 2010, scientists writing in the well-regarded *Journal of Zoology* said they observed behavioral changes in toads fifty miles from L'Aquila just before the earthquake. Their observations of interrupted spawning behavior was interpreted as evidence that toad behavior can be a predictive precursor of earthquakes.[55] But by USGS standards, this claim constitutes prediction based on "unusual behavior of animals," and should therefore be considered crank science. Nevertheless, knowing about the history of prediction, pendulum swings, and boundary work, under the right conditions, might this not be reclassified as legitimate precursor research?[56]

According to the USGS website, exact details of the earthquake may have been unpredictable, but "from the perspective of long-term seismic hazard analysis, the L'Aquila earthquake was no surprise."[57] Seismologists work with large data sets to draw probabilistic hazard maps that allow them to forecast—on long time scales—the probability of earthquake activity (see Figure 5.7).

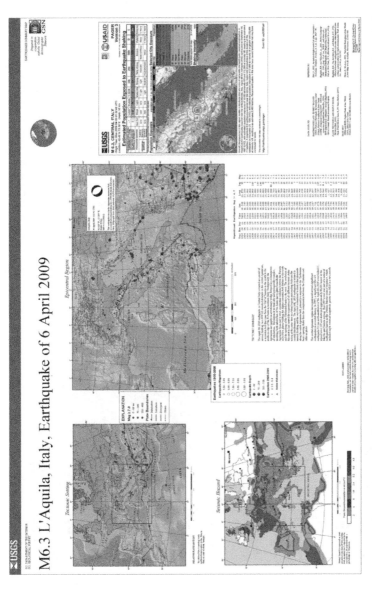

Figure 5.7 Hazard map for central Italy; based on historical seismic data. (Source: USGS; Earthquake Hazards Program Summary Poster. https://earthquake.usgs.gov/archive/product/poster/2009/0406/us/1462221311623/poster.jpg.)

L'Aquila lies within a 30-km-wide seismically active band thread-
ing through the Central Apennines (Figure 5.8). It has a long history of seismic swarms, tremors, and earthquakes.[58] Long-term
statistical forecasting models showed that, sooner or later, a strong
earthquake would strike. The last catastrophic seismic event in the
region was the 1915 earthquake that ravaged the city of Avezzano,
leaving about 34,000 dead.[59] The statistical forecasts and seismic
history created a charged environment around L'Aquila when a
seismic swarm began in the region in October 2008.

For seismologists, such swarms are fairly unremarkable.
Statistically speaking, and over the long term, in a seismically
active region like L'Aquila, for every large-magnitude 6 event, ten
thousand smaller seismic events will be recorded with a magnitude of 2.[60] These seismic swarms do not predict an impending
large-scale earthquake. Swarms can last for days or even months.
They may be accompanied by multiple small shocks, one large
mainshock, or no mainshock at all. In L'Aquila, there was a main-

Figure 5.8 L'Aquila lies nestled among the Apennines in central Italy.
(Source: Photograph by Lasagnolo9 at Italian Wikipedia, CC BY-SA 4.0.
https://commons.wikimedia.org/w/index.php?curid=91152191.)

shock, and it struck April 6. It was preceded by months of tremors and was followed by hundreds of smaller seismic events. All told, the 2008–2009 swarm lasted nine months and included thousands of shocks.[61] Most of these events were all but imperceptible to L'Aquila residents, but tremors occurred often enough that people began to feel anxious about the future.[62]

Media Operations: Local Declarations and Distortions

By early 2009, the seismic swarm had affected the region around L'Aquila for weeks. Then in mid-January a string of small but noticeable shocks struck.[63] Residents' reactions varied. Some called emergency numbers and began to evacuate, but many others assumed this was ordinary seismic activity and carried on without much alarm. Following this string of seismic activity, authorities began trying to get ahead of public anxiety by using the local press to reassure citizens.[64]

Il Centro, a newspaper serving residents in and around L'Aquila, became a key actor in the controversy. Regular articles from January to April described the evolving seismic swarm in real time, thereby reinforcing and sometimes amplifying local anxiety about an earthquake. At the same time, government authorities and scientific experts used *Il Centro* as a platform to tamp down public anxieties. Like a constant drumbeat, these communications drove home specific, anesthetizing rhetorical objectives.

Although such interventions in local media might be understood as national experts' providing objective scientific knowledge, their language suggests a deliberate public relations effort. This was a coordinated media campaign: public officials tried to control the narrative by diffusing anxieties driven by persistent tremors. Anxiety was also fueled by unauthorized predictions of an impending earthquake.

Between January and April, experts were quoted extensively in *Il Centro*. Three main messages emerged. First, activity associated with the seismic swarm was "absolutely not alarming" given that small-scale seismic activity was continuously monitored by sophisticated instruments. Second, the seismic swarm could actually be beneficial as it would "discharge" seismic energy, thereby making a large-scale earthquake less likely. Third, the experts repeatedly stated that "prediction is impossible," thereby explicitly countering the rogue predictions. This campaign of reassurance and demarcation left little room to inform the public about credible, nuanced risk analysis and to reinforce best practices for public safety.[65]

For example, on January 25, Alberto Michelini of the National Institute of Geophysics and Volcanology (INGV)[66] asserted that "currently it is impossible to predict earthquakes but recent earthquakes that have [been felt] can hardly be considered an alarm bell. People should not have any fear."[67] On January 31, four small shocks struck the area, but again experts reassured residents that "there is nothing alarming in this earthquake swarm." Through February and March, on the heels of each series of shocks, more such statements appeared in *Il Centro*.

Prediction was framed as impossible. Therefore anyone claiming to predict was not aligned with accepted scientific practice, and should be considered pseudoscientific—not to be listened to, let alone trusted. To this end, historical and scientific consensus was filtered out. In no public proclamation was there an attempt to place current seismic prediction in historical context. There was no reference to the decades of work by seismologists before the fairly recent prohibition on prediction. Such omissions were reinforced by a key scientific distortion that was part of the media campaign.

On February 18, an expert from the INGV took things a step beyond scientific consensus by claiming that "it's better for the energy to be released a little at a time, rather than all together."[68] The authorities made this point repeatedly, in different contexts. This was the argument, it will be recalled, that the surgeon Vittorini used to convince his wife that they were safe.[69] As many seismologists have noted since, this "discharge of energy" claim is hugely problematic. No solid evidence supports it, and it gives people a false sense of security.[70] Even so, experts gave it out to the press as accepted scientific knowledge. As we shall see in the following sections, this claim, coupled with the other anesthetizing proclamations by scientists and public officials, lulled residents into ignoring traditional practices; in turn, this caused many to feel "betrayed" by science.[71]

Accused Pseudoscientist Giampaolo Giuliani—Predicting Earthquakes

Though the swarm itself was seismically unremarkable, the public controversy over seismic prediction before the L'Aquila earthquake was not. In part, controversy sparked when a local resident, Giampaolo Giuliani, made a series of predictions about impending shocks. With no formal graduate training or degrees, Giuliani was not recognized as a professional scientist. Instead, he was a scientific technician who had spent decades in national laboratories building and maintaining complex equipment. Among other projects, Giuliani had been part of the large volume detector (LVD) team working at the Gran Sasso National Laboratory deep in the Apennines, seeking to detect neutrinos from core-collapse supernovae.[72]

In addition to the LVD work, the Gran Sasso lab houses many other projects in nuclear physics. One that caught Giuliani's attention was the environmental radioactivity monitoring for earth sciences (ERMES) initiative. The ERMES team monitors naturally occurring levels of radioactive isotopes such as radon (^{222}Rn) and uranium (^{238}U) in groundwater running through the aquifer system.[73] ERMES generates a data set that allows scientists to study correlations between seismic activity and the release of radioisotopes dating back decades. Radon received significant attention in these studies, because it is a chemically inert gas with a relatively short half-life (3.8 days). In theory, radon's variations in concentration levels, in groundwater or in the air, can be correlated with specific seismic events.

In fact, credible scientific research studying the release of radon gas as a precursor to earthquakes has been active since the 1960s.[74] For decades, scientists working on this line of research had the explicit goal of establishing a field of "earthquake prediction studies."[75] However, in line with Hough's pendulum swings, by the early 2000s seismologists were saying that regardless of correlations between radon gas and seismic activity, too much uncertainty surrounded the underlying geodynamics to construct a reliable model for earthquake prediction. Scientists continued to gather data on correlations between seismic activity and environmental precursors, including anomalous radon release, but they tended to purge any reference to "prediction" from their professional and public discourse.

Guiliani, however, was intrigued. Dissatisfied with science's lack of progress toward a predictive model of earthquakes, he spent the first decade of the 2000s developing his own, homemade, privately funded, distributed radon gamma-ray detection system. His research was initially inspired by colleagues at the Gran Sasso laboratory, squarely within the boundaries of mainstream accepted science.[76] But Giuliani insisted on going beyond correlation by trying to create a system for earthquake prediction, which damaged his scientific credibility and made funding difficult. This was particularly true because during this time most seismologists were steering clear of predictive models—Hough's pendulum swing again. For years, Giuliani tried convincing seismologists that his research deserved support. He presented his work in various official forums, but always felt rebuffed by the scientific gatekeepers.[77]

Privately, then, Giuliani and a few collaborators continued to develop his radon detection system. By autumn 2008 he had begun to make a name for himself in the local press near L'Aquila. Giuliani was born and lived in L'Aquila, so he had a personal interest in the area. In an October 2008 magazine article, Giuliani claimed to have predicted the 2002 earthquake at Molise, where twenty-one children died when the roof of their nursery school collapsed. Unfortunately, according to Giuliani, in 2002 his system had been in a highly experimental state, so his predictions were not taken seriously.[78] Giuliani also claimed that his now-patented PM4 system had recently been endorsed by the scientific community and rigorously tested as part of a campaign sponsored by the Department of Engineering at the University of L'Aquila. The cost of his system was 60,000 euros,[79] a relatively modest amount compared to the billions of euros lost during earthquakes. For months before the April 2009 mainshock, in the midst of the ongoing seismic swarm, Giuliani took the opportunity to showcase and possibly sell his earthquake prediction system.

As the seismic swarm persisted month after month, Giuliani, convinced he was seeing correlations between radon levels and seismic tremors, began posting "real-time measurements from his detectors" on a public website. He also made specific predictions about future seismic shocks.[80] The first of this series of predictions came on February 17, 2009, and was quickly rejected as a false alarm by seismologists.[81] Then, on March 27, Giuliani warned the mayor of L'Aquila that radon measurements indicated that the town may face a significant tremor within twenty-four hours. Although there were tremors in L'Aquila the following day, they were "imperceptibly small."[82]

Two days later, a more dangerous threat was announced. Based on detections of radon, Giuliani concluded that the city of Sulmona, fifty-five kilometers southeast of L'Aquila, faced an imminent "catastrophic" earthquake.[83] He contacted the mayor of Sulmona, who raised the alarm. Vans with loudspeakers reportedly drove through the city warning residents to evacuate, and priests removed valuable relics from their churches.[84] No noticeable seismic activity followed.

The panic caused by Giuliani's false-positive prediction in Sulmona was the tipping point for government officials, who thought his string of predictions were themselves a threat to public safety. Guido Bertolaso, director of the Department of Civil Protection (see Figure 5.9), began legal proceedings to silence him.[85] Giuliani would be prosecuted for "unnecessarily disturbing the peace"; punitive damages would be sought.[86] A day after the Sulmona false alarm, a legal injunction forbade him from publicly announcing his seismic predictions and required him to remove such warnings from his website.[87]

Although Giuliani's inquiry into seismic prediction was born of legitimate scientific pursuits, he had stepped beyond the accepted scientific consensus of the time and caused public alarm. He was therefore cast out; silenced. Scientists and public officials made no attempt to place Giuliani or his work in its proper scientific or historical context. There was no acknowledgment that he was trained as a scientific technician who for decades had successfully built instruments in prestigious national laboratories, or that he was working on a project that ten years earlier would not have been automatically dubbed pseudoscience by mainstream seismologists.

Figure 5.9 Dr. Guido Bertolaso (Source: Photograph by Roberto Ferrari; Flickr.com, CC BY-SA 2.0, https://commons.wikimedia.org/w/index .php?curid=47529030.)

Performing Certainty: Discrediting an "Imbecile" and Anesthetizing the Aquilani

Outraged by Giuliani's false alarm in Sulmona, Bertolaso convened an emergency meeting of the Commissione Grandi Rischi (CGR), to be held in L'Aquila on March 31.[88] Appointed by the DPC, the Commission of Major Risks is an advisory group of twenty-one technical experts and public officials tasked with providing recommendations about all serious risks to public safety. The regular bimonthly meetings of the CGR typically convened in Rome behind closed doors. However, in light of rising anxiety, Bertolaso felt that the L'Aquila situation warranted a special onsite session of the CGR, open to local officials.[89]

The meeting was officially convened to closely examine "the seismic events of the past four months in L'Aquila and its surrounding area, from the perspective of scientists and civil protection agencies."[90] However, two important agendas were hidden. First was an intent to discredit and stigmatize Giuliani by performing a public act of boundary work leading to a clear demarcation between science and pseudoscience. The DPC felt that Giuliani's rhetorical certainty must be undercut with the unambiguous message that predicting earthquakes is impossible; therefore anyone engaging in predictive modeling should be considered an untrustworthy crank, a pseudoscientist. The second hidden agenda was to use the expertise of the CGR to create a reassuring narrative that would counter Giuliani's predictions. The anesthetizing refrain to emerge from the meeting was the current seismic swarm was completely normal; there was nothing to fear.

As Bernardo De Bernardinis, deputy director of the DPC, brought the CGR to order on the evening of March 31, 2009, he reiterated its public intent: the group had been convened "with the objective of putting together the facts needed to inform the *Aquilani* about the seismic activity of the past few weeks."[91] That was a sensible, fairly typical objective in light of the CGR's mission. Yet this meeting was far from typical. It was not simply a fact-finding policy meeting. The intent was not solely to have experts debate scientific findings and come to a consensus that could then be used to inform the *Aquilani* about risk. They could have done that in Rome. Instead, Bertolaso was intent on building a public stage upon which CGR experts could perform acts of demarcation and anesthetization.

This may appear to be an overly dramatic, even sensational, characterization of the CGR meeting. But during the legal and social firestorm that followed the L'Aquila earthquake, Bertolaso's previously hidden agendas of demarcation and anesthetization were fully uncovered. In January 2012, police released transcripts of phone conversations between Bertolaso and one of his deputies at the DPC, recorded a day before the CGR meeting of March 31, 2009. The recording was obtained as part of an unrelated corruption investigation, but the implications for the L'Aquila Seven, the DPC, and the CGR were unmistakable.[92]

In that March 30 phone call, Bertolaso explained to his deputy that he was convening the emergency CGR meeting so he could send "the leading earthquake experts" to L'Aquila as part of a "media operation" that would "immediately shut up any imbecile and calm down conjectures, worries, and so on."[93] Bertolaso added, "We are doing this not because we are worried, but because we want to reassure people."[94] The transcript makes it obvious that the CGR was not focused solely on engaging in meaningful scientific analysis of risk in L'Aquila. The group had a mixed agenda that included performing the last act of an ongoing media operation to anesthetize the *Aquilani* and discredit the imbecile in question, Giuliani.

Boundary Work within the L'Aquila Seven—Refracting Uncertainty

In the scramble to assign blame, another instance of boundary work became visible. This time the boundary was not between scientists vs. pseudoscientists or prediction vs. forecasting. Rather, it was a pragmatic act of "protective amputation." Most analyses make a point of the group's composition: the Seven are partitioned into two subgroups. Six are labeled scientists and assigned expert status, while the seventh is introduced as a government (or public) official. Because expertise is at the heart of this case, it is worth examining the partition more carefully.

The six scientists are the following: (a) Franco Barberi, a volcanologist at the University of "Rome Tre" and chair of the CGR; (b) Enzo Boschi, geophysicist and president of the National Institute of Geophysics and Volcanology (INGV); (c) Gian Michele Calvi, seismic engineer and president of European Center for Training and Research in Earthquake Engineering; (d) Mauro Dolce, engineer and head of the DPC's seismic-risk office; (e) Claudio Eva, physicist at the University of Genova; and (f) Giulio Selvaggi, seismologist and president of the INGV's National Earthquake Center. The lone "public official" is Bernardo De Bernardinis, deputy director of the DPC.[95] Although his official title is sometimes included in examinations of the controversy, De Bernardinis's background as an academic with a PhD in hydraulic engineering and a long list of prestigious scientific appointments is not.[96]

Looking at the L'Aquila Seven and their various areas of expertise, one can imagine many ways of partitioning the group. In terms of training, three are engineers and four natural scientists. Three have training in seismology and four do not. Geographically, five are based in Rome; two in other Italian cities. No partitioning scheme yields consistent groupings, and none of these categories seems particularly useful in this context. Why has the particular partitioning of six scientists vs. one public official stuck? Strictly speaking, there are not six natural scientists in the Seven, and the one "public official" is an engineer. If one were grouping engineers and natural scientists together, all of the L'Aquila Seven would be classified as scientists. In any event, the standard categorization is debatable.

It turns out that rhetorical reasons prompted this particular demarcation. First, the scientist/public official binary blurs the distinction between natural scientists and engineers. Regardless of their specific training, each contributed their expertise to the CGR deliberations and is part of accepted scientific discourse. The blurring produces an apparent cohesion among the six scientific experts and outsider status for De Bernardinis, even though he, too, was an expert in hydraulic engineering.

This rhetorical strategy is often seen in scientific myth-histories. In order to protect its myth-historical imaginary, the community disavows some of its members, stigmatizing them as bad actors. Labeling De Bernardinis a public official without mentioning his technical expertise rhetorically strips him of scientific credentials, culling him from the herd of scientific experts.

Why did De Bernardinis become the scapegoat? As deputy director of the DPC, he worked closely with Bertolaso to orchestrate the "media operation" that discredited Giuliani and anesthetized the *Aquilani*. Although the other members of the Seven also participated in this operation with varying degrees of awareness, they were not directly involved in planning it. Indeed, the minutes of the March 31 meeting show distinct phases in the rhetorical use of uncertainty. The shifting language reveals differences in assumptions and desired outcomes among participants.[97]

The first phase of the CGR meeting cleaves to its stated mission. The scientists matter-of-factly presented historic and more recent seismological data on the L'Aquila region. In "objectively" reporting their observations as facts, they tended to use "is" statements to stress unequivocal confidence in their findings. However, as these same scientists switched gears and began discussing what the data meant, their language changed, now including qualifiers and other markers of uncertainty.[98] In other words, the scientists were certain about their data but uncertain of what conclusions could be drawn.

As discussion moved on to Giuliani, clear bifurcations emerged between the language of expert scientists and those of policy makers, like De Bernardinis, associated with the DPC. DPC representatives seized scientific uncertainty and argued that Guiliani should be silenced and the *Aquilani* reassured. The scientists agreed that predictions of seismic activity were not possible, but their comments about Giuliani were far more qualified. They did not dispute the data collected at his radon detectors, acknowledging that the fluctuations in radon levels most likely reflected real variations. In addition, they did not discount the possibility that someone might use these as a basis for predictive models of seismic activity "in the future."[99]

The minutes reveal a deep divide between those who acted like scientific experts, strictly in an advisory role, and those who, regardless of scientific expertise, participated in the meeting as policy makers working for the DPC. The two groups used different rhetorics and repeatedly talked past each other, ultimately failing to communicate clearly. This miscommunication betrayed distinct preexisting assumptions and agendas that made the meeting more a performance than a genuine dialogue.

Beyond Scientific Consensus—Anesthetizing Misinformation

Policy makers like De Bernardinis, tasked with communicating CGR findings to the *Aquilani*, failed to represent scientific knowledge and uncertainty accurately—not because they couldn't understand what the scientists presented, but because they had predetermined the outcome. This point became incandescently clear when a TV news station aired an interview with De Bernardinis just after the CGR meeting, an interview recorded well *before* the meeting took place on March 31.

In the interview, De Bernardinis gave out questionable advice and made problematic scientific claims.[100] Indeed, he seemed to be reading from the same script as his boss, Bertolaso. He dismissed the possibility of seismic prediction and downplayed risk to the *Aquilani*. Based on Bertolaso's comments in the phone call of March 30, 2009, and his later testimony during multiple trials, neither he nor De Bernardinis was worried about an earthquake. Both men claimed that the ongoing seismic swarm around L'Aquila was "actually favorable," as it resulted in a "continuous discharge of energy" that would make a catastrophic event less likely.[101]

Bertolaso, De Bernardinis, and other members of the DPC repeatedly made this claim about discharging energy. But it is a hypothesis without evidence. During the trial of 2012, seismologists made this explicit.[102] The scientific community now seems to disavow the discharge of energy idea; they certainly do not claim it as part of their consensual knowledge. But during the media campaign of 2009, no CGR scientific or technical expert stood up to refute the hypothesis. Did silence make them complicit? According to Judge Billi, it did.

The veracity of the discharge of energy claim is less important to this analysis than the fact that Bertolaso and his deputies at the DPC seemed to believe it was a reasonable scientific claim. In that historic phone call, Bertolaso said, "It's better that there are a hundred magnitude-4 tremors rather than silence because a hundred tremors release energy, and there won't ever be the damaging tremor."[103] De Bernardinis was so confident of this claim that, prompted by a reporter's question, he emphasized the discharge of energy hypothesis and suggested that instead of worrying about a major earthquake, the Aquilani should relax, go home, and enjoy a glass of wine.[104]

For Vittorini and those most affected by the earthquake, this flippant exchange between De Bernardinis and the reporter became emblematic of the scientific establishment's abandoning its responsibilities and failing to keep the citizenry safe. In their defense, the six other members of the L'Aquila Seven were quick to expel De Bernardinis from their ranks. According to their lawyers, De Bernardinis alone had communicated the discharge of energy hypothesis, which gave a false sense of security to the public.

The claim is, however, refuted by many articles in *Il Centro*: the discharge hypothesis was mentioned repeatedly in the months leading up to the CGR meeting. In addition, the minutes of this meeting show that when the discharge of energy hypothesis was mentioned, no one challenged its validity. A prosecutor in the appeals case made this point explicit: "Why on 31 March did no one dissent, no one jump up out of their seat, no one explain to the other people present [that the positive effects of the discharge of energy were] nonsense and not a positive signal?"[105] Finally, even after the CGR meeting, when De Bernardinis's television interview was aired nationally, none of the experts stepped forward to correct the misinformation.

This contextual evidence causes problems for Leshner's portrayal of Judge Billi's ruling as an "unfair and naïve" adjudication.[106] The L'Aquila Seven were not just scientists "adhering to accepted scientific practices" and engaging in "the free exchange of ideas."[107] The evidence seems to indicate that De Bernardinis and other members of the DPC were, to varying degrees, complicit in the anesthetizing and demarcating media operation. The other six were initially found guilty because of their work on the CGR and their silence in the face of the DPC's misinformation campaign.

Fallout

Before 2009, there was a general culture of preparedness in the seismically active region of Abruzzo. In the face of seismic activity *Aquilani* traditional knowledge had taught people to spend the night outside on the piazza. However, as the calendar slipped from March to April 2009, the approach to seismic preparedness changed.[108]

Facing greater public anxiety, partly due to Giuliani's rogue predictions, the DPC intensified its media campaign to discredit Giuliani and to anesthetize the citizens. Such boundary work used the scientific authority of the CGR experts and mischaracterized scientific consensus. As a result, local traditional knowledge was overwhelmed by the performed certainty of scientific authority; it became "unnecessary" and "superstitious" to many *Aquilani*.[109]

In the aftermath of the earthquake—and in the social aftershocks that reverberated through the global scientific community—we are left wondering how it could have happened. Hundreds died, but it is difficult to show to what extent the DPC media operation contributed to those deaths. The earthquake itself was an unpredictable, uncontrollable disaster, but the social performance of science clearly played some role, too. A confluence at the intersection of science's uncertain knowledge, scientists' performance of their uncertainty, and the public's understanding of uncertainty created a perfect and fatal storm.

In 2014, an Italian Court of Appeals disagreed with Judge Billi's assessment, overturning his decision. Of the L'Aquila Seven, the six "scientific experts" were acquitted of all charges. De Bernardinis's conviction was upheld, but his sentence was reduced to two years and suspended indefinitely. As the decisions were read aloud, shouts of "Shame, shame!" rained down on the three appellate court judges. For many of the victims' families looking for justice, the decision was an "earthquake after the earthquake."[110] The final event was the acquittal of Guido Bertolaso in 2016. He had been charged separately from the Seven for his role as director and mastermind of the media operation. The court ruled that not enough evidence linked Bertolaso to De Bernardinis's misrepresentation of scientific certainty.[111]

It is beyond the scope of this analysis to decide whether the L'Aquila Seven were all equally culpable of spreading misinformation. It is clear, however, that by discrediting Giuliani and presenting the discharge of energy hypothesis as fact, the DPC was protecting their myth-historical ideals. They overstated scientific certainty for rhetorical ends, doing boundary work that reverberated far beyond their intentions. When the social and legal backlash came, they were again quick to defend their ideals, this time by sacrificing one of their own. In this second round of demarcation,

De Bernardinis was stripped of his scientific credentials and labeled a "public official." He was a tumor to be excised, not allowed to fester and ruin the reputation of his peers or stain science's good name.

Conclusion

When we discuss scientists behaving in ways counter to scientific ideals, why do we label them corrupt or incompetent? This type of boundary work is part of scientific practice. It helps define inclusive and exclusive social circles, and reinforces science's commitment to protect its myth-historical ideals. The dissonance between ideal science and practiced science is the tension felt in Einstein's two divergent characterizations. Given the "reproducibility crisis" and the extent to which it has become a problem in many disciplines, we can no longer claim that the scientific community as a whole behaves ideally while singular scientists fail to live up to our stated norms.[112] Conscious or not, activities like selection bias and p-hacking are rampant and have serious implications for the science that is produced.[113]

One might want to claim that in the long run science is self-correcting: any idiosyncrasies or distortions in the short term will eventually be filtered out. However, what if science's corrective filters are also psychologically conditioned and less effective than we think? Or what if the corrective filters work, but on time scales that seem unbearably long? Public trust can suffer irreparable harm in the meantime. Should scientists trust this self-corrective mechanism, or should they take a more proactive approach to shaping their scientific imaginaries?

In *Merchants of Doubt*, Naomi Oreskes and Eric Conway uncovered a network of actors who raised doubt in several important controversies. From the effects of smoking to climate change, the authors tracked the alliances of business, lobbyists, politicians, media, and scientists who promoted their agendas even when they conflicted with scientific consensus and the greater good.[114] Oreskes and Conway showed what could be gained by obfuscating scientific consensus. A handful of scientists with political agendas can fuel misinformation campaigns that undermine public trust in science.

In the broader context of a global "war on science," we can ask how scientist-storytellers contribute to a misunderstanding of science and a corresponding public mistrust. After the L'Aquila earthquake, many *Aquilani* clearly felt betrayed by science. Although betrayal does not normally have life-or-death consequences, even when the stakes are lower, rhetorical distortion can damage the public's image of science. There is inherent tension between how scientific communities deal with controversy and how the public perceives them. This reality must be at the core of any analysis of a "war on science."

Carl Sagan noted that humanity has become profoundly dependent on—and simultaneously ignorant of—the workings of science and technology.[115] I agree with Sagan, but he fails to recognize that he and other scientist-storytellers are partly responsible for the dissonance. For too long their myth-historical portrayals of progress have championed the erroneous view that science is the only true candle lighting the darkness of ignorance. By focusing on past successes and filtering out failures, scientific narratives leave audiences with the impression that science is associated only with rational truth and certainty.

Myth-histories describe scientists as heroes committed to rigorous methods, which permit them to read the book of nature and unveil the great mysteries of the universe. Rarely are scientists described as grappling with insurmountable uncertainty. The evolution of scientific knowledge is presented as inevitable, with little room to explore uncertainty in scientific practice. The scientific ideal of certainty has become a myth-historical trope to be protected at all costs. Yet sometimes the costs are high. The L'Aquila earthquake illustrates how important it is for scientists to be aware of how they manage uncertainty. When people feel in partnership with science, they tend to be more understanding of scientific uncertainty. When they are intentionally excluded from scientific practice and treated paternalistically, the seeds of mistrust are sown.

Conclusion-Beyond the Wars on Science

Stories matter. Many stories matter. Stories have been used to dispossess and to malign, but stories can also be used to empower and to humanize.[1]

—CHIMAMANDA NGOZI ADICHIE

The Flying Ashtray Revisited

In Chapter 1 we examined the encounter that Errol Morris dubbed Thomas Kuhn's "ashtray argument." Morris's memoir reveals much more than an alleged assault. His book aims to discredit Kuhn's ideas and to question their legacy in both popular and academic cultures. He considers Kuhn's *Structure of Scientific Revolutions* (*SSR*), at best, "an inchoate, unholy mixture of the work of others," and at worst, "an assault on truth and progress."[2] According to Morris, Kuhn's ideas "promote a [dangerous] denial of truth," and are partly to blame for recent trends toward a "devaluation of scientific truth."[3]

This contrasts sharply with what Kuhn himself intended, and significantly diverges from most serious readings of his celebrated work.[4] Kuhn and his legacy are Morris's proverbial "ax to grind."[5] His book is a caustic, quixotic charge against a caricature of Kuhn's ideas, using selected quotes to vilify. For example, according to Morris, Kuhn's "tripartite conviction" stands for "No progress. No objective truth. No real world."[6] Yet Kuhn, trained as a theoretical physicist, repeatedly stated his unequivocal belief in both progress and the real world. His views on scientific truth were far more subtle than Morris's distortion suggests.[7]

One almost hears Kuhn, from beyond the grave, yelling "I am not a Kuhnian!"[8] Kuhn's legacy has become "all things to all people."[9] Part of the responsibility for this multiplicity of (mis)representations falls squarely on Kuhn. His language was at times provocative and slippery. After a revolution, are scientists really working in a different world, or do they perceive their world differently? Did Kuhn really mean that scientists in successive paradigms could not translate their work at all, or is his use of "incommensurability" somehow qualified and less sweeping?[10]

Interpreters also bear responsibility for diverging interpretations. *SSR* is the quintessential accidental academic bestseller. Written as an essay for the *Encyclopedia of Unified Science*, it was intended only as a "schematic...first presentation"; a stepping stone to a "full-scale book" that never materialized.[11] Yet its sometimes abstruse content has been devoured, interpreted, and (mis)represented by an endless line of academics and popularizers using varied frameworks.[12] Although Kuhn's proposed structure in *SSR* was never meant to be a magnum opus, many critics have dissected it as just that. Kuhn considered *SSR* a first pass at explaining what makes science such a powerful human phenomenon, and to simultaneously correct Whiggishly distorted portrayals of its history. As a practicing historian of science, he relied on careful analysis of historical case studies to make his arguments, believing that doing so might ultimately lead to a more faithful representation of science and its evolution. Kuhn's iconic opening sentence in *SSR* lays out his methodology and intentions: "History, if viewed as a repository for more than anecdote and chronology, could produce a decisive transformation in the image of science by which we are now possessed."[13]

Much has changed since Kuhn wrote those famous lines in 1962, yet his claim haunts science today. We must take history seriously, now more than ever. If we really believe Shawn Otto's claim that there is a "war on science," or Errol Morris's contention that scientific truth is being devalued, we should look beyond the causes they identify: boogeymen postmodernists, social constructivists, and Kuhnian disciples.[14] Such polemical characterizations lack nuance, echoing the caustic science wars of the mid-1990s. This book has sought to highlight problems in the rhetoric of observers like Otto and Morris. By exploring the impact of the myth-historical stories they tell, we can encourage scientists to reflect on their own role in these controversies.

Reflective Discourse

A key argument of this book is that we need more thoughtful and reflective interdisciplinary conversations on the topic. I support Naomi Oreskes's suggestion that a more transparent "articulation of [the] social character" of science is needed.[15] Her book *Why Trust Science?* explores problems related to Morris's devaluation of scientific truth, but from a fresh vantage point. Her argument centers on the claim that the power of scientific knowledge stems from its "fundamentally consensual" nature.[16] However, it turns out that unpacking the notion of consensus can cut both ways when it comes to trust in science. On one hand, understanding how scientific consensus is formed reveals the underlying social fabric of which science is constructed. During the 1970s and 1980s, early science studies scholars used this understanding to challenge scientific authority and question the widely accepted objectivity of scientific knowledge.[17] Their provocative studies of the social underpinnings of science uncovered subjective sources of bias that were absent from traditional communal imaginaries.[18]

These imaginaries were informed by the myth-historical stories scientists told about their craft. As we have seen, such narratives tend to filter out social elements that were inconvenient or undermined scientific norms and ideals. The growing dissonance between polished tales of idealized science and the messy evidence of science in action became a lightning rod for confrontation. This friction was most notable in polemics during the mid-1990s science wars. Had scientists been more reflective and transparent about the underlying social dimensions of their craft, and scholars of science studies less intentionally provocative in their relativistic ontological claims, the dissonance could have created a healthy interdisciplinary conversation rather than hyped-up polemics. Perhaps this period resembles the current state of devalued scientific truth and eroded trust. One strategy to address these problems of dissonance is greater transparency.

An honest accounting of the "social character" of science does not reflect weakness in science, but its greatest strength. The social underbelly of science actually provides the strongest case for scientific objectivity.[19] As feminist philosophers of science have argued for decades, "Diversity serves epistemic goals."[20] A community of scientists who have different life experiences ensures heterogeneity in "standpoints" and interpretations of the world.[21] In turn, diversity becomes one of the best means to test and correct scientific knowledge. In studying science's methods of self-correction, philosopher of science Helen Longino noted that self-correction isn't magical or spontaneous; scientists "correct *each other*" through social processes that constitute "transformative interrogation."[22] The "depth and scope" of this social interrogation determines the objectivity of the scientific knowledge produced.[23]

In theory, if a heterogenous group of individual experts from divergent standpoints work on a particular problem and come to the same conclusion, it ensures that their consensus understanding is more objective. Their scientific conclusion is less likely to be informed by any single ideological bias. This vision differs from the Mertonian ideal of universalism. Traditional universalism assumes that scientific knowledge is true irrespective of a person's standpoint; diversity is irrelevant to objectivity. Myth-historians have traditionally projected this ideal by filtering out social dynamics and personal perspectives from their representations of history.

By contrast, standpoint theory claims that scientific knowledge is more objectively true and trustworthy *because of* science's social character, not despite it. If so, then scientists must be more transparent about their social engagements. Scientists must become better communicators, explaining "not just what they know, but how they know it."[24] If scholars in other fields, students, and the public had a better sense of how scientists do research and engage each other in various forms of social "transformative interrogation," they would be more apt to trust them. This point is illustrated by a recent episode of grand scientific claims, a premature press conference, and a perceived signal that turned out to be cosmic dust.

Transformative Interrogation in Action

Almost two years before the LIGO press conference discussed in Chapter 4, there was a dress rehearsal, of sorts, with a different cast of scientists and dramatically different outcomes. While the LIGO team's announcement led to Nobel Prizes for its lead scientists, leaders of this earlier team found themselves watching a different drama, as their Nobel dreams "bit the dust."[25] On March 17, 2014, scientists from the BICEP2 collaboration gathered at the Harvard-Smithsonian Center for Astrophysics (CfA) in Cambridge, Massachusetts, for a press conference to announce a monumental discovery.[26] Unlike the LIGO announcement, the BICEP2 team did not claim a direct detection of gravitational waves; they were announcing the first "images" of gravitational waves. However, this imaging promised a new window into the "first tremors of the Big Bang," and, as a result, the "first direct evidence" for "cosmic inflation."[27]

The BICEP2 telescope at the South Pole had collected data for three years while hunting a special type of swirly-like "B-mode polarization" imprinted ever so slightly on the cosmic microwave background (CMB) radiation.[28] The CMB is the closest thing we have to a baby picture of the universe. For the first 380,000 years, expanding remnants of the Big Bang were so hot, dense, and energetic that electromagnetic radiation could not escape. As a result, the earliest universe will always be hidden behind an opaque veil with no photons available to tell us what it looked like. It is hard to overstate the importance of this veiled CMB window. Many of the most fundamental discoveries in cosmology over the past two decades have come from analyzing tiny fluctuations in temperature and density in the CMB. In 2014, the BICEP2 team was the latest in a long line of research teams struggling to decipher the mysteries of the early universe by studying the CMB.

The BICEP2 collaboration was trying to isolate the B-mode signal within the CMB to gather observational data that would support inflationary models of the early cosmos and restrict the parameters of their viability. In these models, the "bang" in Big Bang refers to an infinitesimal moment within the first fraction of a second in which the universe rapidly inflated, expanding exponentially.[29] Over the past few decades, inflationary models have challenged our conventional understanding of space and time, evolving from highly speculative ideas to mainstream theoretical building blocks of today's standard cosmological paradigm. Isolating the B-mode polarization associated with primordial gravitational waves would be like finding "smoking gun" evidence for cosmological inflation, and would help place robust limits on viable models. It would clarify our understanding of the early universe and could yield prizes and praise for the scientists responsible.

The ripples in what Harry Collins calls "social spacetime" after this announcement illustrate the contingent and uncertain nature of scientific practice. Apart from the people physically present at the press conference, countless people from around the world had tuned in virtually. News of the discovery made the front pages of many newspapers. Headlines relayed the confidence and certainty communicated at the press conference: "Big Bang's 'Smoking Gun' Confirms Early Universe's Exponential Growth," said one.[30] No hesitation, no contingency. The problem was that the BICEP2 team went public before their work was properly reviewed—that is, peer reviewed.

Marc Kamionkowski, the unaffiliated expert in B-mode polarization signals, did sound a warning: "We must wait before buying any tickets to Stockholm…or to reach these dramatic conclusions until these results are vetted fully by the community and verified by other groups."[31] But while other cosmologists were diligently checking BICEP2's findings, the public hype machine was grinding. To be sure, the dozens of scientists participating in the BICEP2 collaboration had been scrutinizing their data for years, running every type of calibration and cross-check they could think of. This was a collaboration of experts from elite institutions, who did their due diligence and genuinely thought they had discovered the ephemeral B-mode polarization signals.

The group was aware that the signals detected could be due to interstellar dust. Nevertheless, using data from their own telescope and what they could glean from a competing group's online presentation, they convinced themselves that the signals were due to primordial gravitational waves, not dust. But they had misinterpreted the other group's data, and relied on it too heavily in their statistical modeling. After careful peer review by other scientists, the BICEP2 claims were declared untenable. Eventually their cherished, much-hyped, B-mode polarization signals were shown to be artifacts of interstellar dust, not primordial gravitational waves.

The saga of BICEP2's claims and ultimate rejection shows "transformative interrogation" in action. The BICEP2 team initially erred; the community peer reviewed their work and corrected the errors. Even more encouraging, the competing group from the European Space Agency (ESA) collaborated with BICEP2 on a joint analysis to scrutinize each other's data. As a result, there was no protracted controversy.[32] Since 2014, the BICEP team has developed increasingly sensitive instruments at the South Pole in pursuit of those elusive primordial gravitational waves. Invoked by some as a cautionary tale against prematurely announcing scientific results, the BICEP2 episode can also be seen as a triumph of consensual science.

But how is the public to interpret this controversy? If they don't understand what happens in transformative interrogation, how will they read flip-flopping headlines? Astrophysicist David Spergel was asked how the BICEP2 controversy might affect the public's perception of science: "[The] optimistic scenario is that the public will say, hey, this is great, it helps science when people check each other's data." On the other hand, the public might interpret this episode as an indictment of science. Spergel continued: "As one of my right-wing Facebook friends remarked: If the scientists get gravity waves wrong, why should we believe them on global warming?"[33]

We can't know a priori how the public will react. Either way, it's important for scientist-storytellers to realize that they have a critical role to play. For example, Brian Keating, a member of the BICEP collaboration, published a forthright, accessible essay about the 2014 episode.[34] In "How My Nobel Dream Bit the Dust," Keating did not hold back; he delivered a thoughtful critique of his team's actions. His storytelling shows vulnerability and humanity, embracing the messiness of science rather than filtering out all the uncomfortable details. More scientist-storytellers need to embrace this transparent approach in their accounts.

Full Disclosure and Transparency Extends to the Past

The potential benefits of a reflective approach to storytelling is not limited to scientists' communications about contemporary research. Scientist-storytellers who tell myth-histories for pedagogical and other rhetorical purposes would do well to consider the implications of their stories. Myth-historical scientific imaginaries like the one described by Leon Lederman can be rhetorically useful in the short term, but the long-term repercussions can be uncertain. Like ripples in social spacetime, dissonance between imaginaries and reality may contribute to an erosion of credibility.

In Chapter 2 we saw that quantum myth-history was consoli-
dated by the likes of Heisenberg and Born, winners of the interpretation debates of the 1920s. Accordingly, J.S. Bell, who began
studying quantum mechanics in the late 1940s, was told only of
one "correct" interpretation: the Copenhagen interpretation.
Alternative interpretations, like de Broglie's pilot wave theory,
were considered dead ends: abandoned, forgotten, filtered from
myth-histories. This act of omission was not an act of villainy.
Quantum storytellers were pragmatists; they told stories they
thought would help students focus on the correct problems and
avoid unnecessary wrong turns. Yet the effects of such omissions
were real. The Copenhagen interpretation galvanized consensus,
resulting in decades of fruitful applications and advancements.
But it also lulled interpretation debates and marginalized those
interested in pursuing interpretive questions.

We also saw, in Chapter 4, a community of scientists reverse
decades of myth-historical filtration and turn a pariah into a
hero. The LIGO leadership rehabilitated Weber's reputation. All
mention of the crushing marginalization he experienced is made
invisible, along with his undeniable obstinacy. Given how some
scientists mistreated Weber, one might see this as justifiable res-
toration. After all, what's the harm in celebrating someone, even
at the cost of historical accuracy? Although LIGO's new myth-
history might not have any immediate casualties, beyond historical truth, the cumulative effect of dissonance between the
imaginaries painted by myth-histories and the realities of science
in the making may eventually take a toll.

In the quantum theory and LIGO examples, myth-histories
were not part of intentional misinformation campaigns. They
were born of pragmatism and a sense of justice, as genuine
attempts to rewrite the past in order to move science forward.
However, in some cases—such as Eric Lander's CRISPR narrative,
in Chapter 3, and the Italian scientists' demarcation attempts, in
Chapter 5—scientists' use of myth-histories is more closely tied
to particular political or economic ends. As such, the effects of

myth-histories may be more severe—even deadly, as we saw in the case of the earthquake in L'Aquila. But no matter the intent or context of any given myth-history, scientists need to become more reflective about and responsible for the stories they tell.

Throughout this book, we have seen that scientists' authority does not necessarily extend to the history, philosophy, or sociology of their subject. Even so, I disagree with Steven Shapin's admonition that scientists should not write about their own history.[35] Though myth-histories are not to be confused with authoritative, scholarly histories written by historians, they have social value, if not as factual representations of the past, then at least as integral parts of a communal, coherent social existence. Scientists might ask themselves certain questions before preparing their stories. What do they hope to achieve by telling a story? Does the story present science as messy, contingent, and human, or does it sweep away the social character of science in favor of a caricatured progress narrative? Who is cast as the hero, and who is purged and forgotten?

Final Reflections on the Ongoing COVID-19 Pandemic

It has been almost fifty years since Kuhn's ashtray hit the floor. Yet myth-historical narratives that retain a Whiggish presentism are as prevalent as ever. It's as if scientists believe that the communication of science requires dressing it in an idealized costume that hides social conditioning behind a veil of scientific imaginaries. As I write these final thoughts about myth-history, we are in the midst of a pandemic of historic proportions. The COVID-19 crisis is a quagmire, worsened by the repercussions of Morris's devaluation of scientific truth and Oreskes's analysis of the deterioration of scientific trust. In the U.S., the Centers for Disease Control and Prevention (CDC) is regularly dismissed, ignored, and derided by large swaths of the population and even by political leaders.

In this turbulent context, Dr. Anthony Fauci, Director of the National Institute of Allergy and Infectious Diseases (NIAID) and a critical member of the White House Coronavirus Task Force, participated in a podcast that illustrates much of the tension discussed in this book.[36] On June 17, 2020, Fauci was a guest of Michael Caputo on his podcast "Learning Curve." The thirty-six-minute conversation between one of the world's foremost experts on infectious disease and a lay person is remarkably revealing. After some initial pleasantries in which Caputo asked Fauci about growing up in Brooklyn and his early career, there is a fascinating exchange. Caputo claims that Fauci is "always frank" and chooses to "stick to the science," but that this approach ends up confusing people.

Fauci hypothesizes that people might be confused by changing recommendations (e.g., no masks vs. masks) because they don't understand that these policy recommendations are based on an incomplete set of "facts and evidence," evaluated in real time as part of managing an "evolving outbreak." As the situation changes, so does the set of facts and therefore the recommendations. Using examples of personal acquaintances, Caputo then describes a polarized citizenry. Some hang on Fauci's every word; others disbelieve everything he says.

At this point Fauci's tone changes. He highlights what he considers a serious problem in the U.S.—a combination of an "anti-science bias" that is "inconceivable" and a tendency to reject "authority." He decries this as an "unfortunate" combination "because, you know, science is truth. And if you go by the evidence and by the data, you're speaking the truth." In this exchange, Fauci drops all pretense of nuance and subtlety; instead he taps into and reinforces a scientific imaginary. Slipping seamlessly through Einstein's black box, he transforms a highly contingent science into something objective, complete, and equated absolutely with truth.

Caputo responds with a comment that leads to the following exchange:

MC: You know, so it's interesting, doc, because I kind of see that the people who don't believe science are people who believe in absolutes. The truth is it's either true or it's not.

AF: Right.

MC: And in this process, we've seen the models shift. We've seen the data shift. We've seen an instruction shift. And I think perhaps those who believe in absolute truth don't really end up believing science that shifts. Don't you think that in the end, the American people have to begin to understand that science is an absolute truth?

AF: Right.

MC: It really isn't.

AF: Well, science has a—no, I think we have to be careful we don't confuse people. So, let me take a different perspective, Michael.

MC: I AM HERE TO CONFUSE PEOPLE.

AF: Okay. [laughter]

In this exchange, Caputo observes that people distrust science because, while they believe in certainty and absolutes, in the case of COVID-19, they are experiencing something far less than scientific certainty. They are face-to-face with equivocal science, uncertain science. This is critical to the argument of this book. If the expectation of science, consistent with prevailing imaginaries, is that it has access to some objective Book of Nature that contains certain Truth about our world, then the real practice of uncertain science will *always* fall short. By the end of the exchange, Caputo's interjection clearly sows confusion. What's wrong with that? Humans are messy, social dynamics are messy, so why shouldn't science be messy?

In light of the confusion, one can almost hear Fauci downshift his rhetoric, qualifying his previous comment equating science with truth. Fauci pivots, "Okay. So, science is the attempt in good faith to get to the facts, and it isn't perfect." He goes on to explain that although scientists "in good faith" can sometimes observe and interpret things incorrectly, "the beauty of science is that it's self-correcting." The scientific process involves many scientists asking the same questions but working independently. "Sooner or later, something that really is true, will get confirmed time after time, after time."

Fauci concludes with a critical point: "As long as science is humble enough and open enough and transparent enough … to accept the self-correction, it's a beautiful process." Even in qualifying the scientific ideal, Fauci ensures his modified imaginary remains mostly intact. As countless scientist-storytellers have done before him, Fauci deftly navigates the tension underpinning the scientific ideal. Acknowledging the psychologically conditioned imperfections of individual scientists, he reaffirms the sanctity of the collectively imagined community that is science.

The tactic seems to work. After listening to Fauci, Caputo claims to be "science stunted," a problem he shares with "most Americans." Yet, in listening to Fauci's description, Caputo seems to have captured the scientific ideals that scientists seem intent on communicating. "I understand that science is kind of an iterative process," by which one "eventually" arrives at "absolute truth." He then observes that most Americans don't understand this characterization of science as a dynamic self-correcting process that ends up with "one immutable truth." Fauci's succinct response to this is "Right." Based on his previous qualifications of the scientific ideal, Fauci could have—should have—corrected Caputo's use of "absolute truth" and "one immutable truth." Instead, he let it slide. Just like that, Caputo's intervention questioning Fauci's equivalence of science with truth is quickly filtered out of the discourse.

Near the end of the podcast, Caputo asks, "How important is transparency along the way to getting the American people to believe in what the ultimate result is?" To which Fauci responds: "It's everything. And if you don't have transparency, you don't have confidence. If you don't have confidence, you're a dead duck." This is perceptive. In telling a story of science's role in managing the pandemic, Fauci had a unique opportunity to be more transparent about scientists' messy struggle with uncertainty. After defaulting to the opaque imaginary science = truth, he quickly shifted to a slightly more transparent description that admitted imperfections and self-correction. Yet instead of pursuing this line and being explicit about the social processes that constitute "transformative interrogation,"[37] he stopped short, labeling science a "beautiful" process, and allowing Caputo to default back to the original scientific imaginary.

If there truly is a "war on science," the battle of COVID-19 may prove to be a turning point. The preservation of scientists' authority is contingent on ending the dissonance between ideal projections and actual scientific practice. One way to do this would be to legislate that all scientists behave strictly according to prevailing socially codified ideals. A more realistic approach would be to roll up our sleeves and take control of the existing imaginary. Imaginaries are necessary, but we are not bound by them. There is power in realizing that they dynamically evolve, because then we can revise them—bring them out from behind the curtain. The myth-histories scientists tell are not always pernicious; what is pernicious is a fixation on a static, simplistic, and homogenously opaque scientific imaginary. "Stories matter. Many stories matter."[38] It is time to abandon unreflective sleepwalking in scientific storytelling. Let us use our stories to humanize science.

Notes

Introduction: Reconstructing Scientific Pasts

1. Muriel Rukeyser, "The Speed of Darkness," in *The Collected Poems of Muriel Rukeyser*, ed. Janet E. Kaufman and Anne F. Herzog with Jan Heller Levi (Pittsburgh, PA: University of Pittsburgh Press, 2005). "The Speed of Darkness" (1968); Kindle edition, Subsection 5.
2. Pierre Hohenberg died on December 15, 2017. I considered him a mentor, sparring companion, and most of all, a family friend.
3. After years, the moment remains seared in my memory. Although I can't guarantee that this is an exact quote, it's close, and I feel certain it captures the message he intended to get across.
4. John Gribbin, *In Search of Schrödinger's Cat: Quantum Physics and Reality* (New York: Bantam Books, 1984). As an aside, I still remember Joe Eberly's course fondly. Although it was only the second physics course I enrolled in at the University of Rochester, it was definitely one of the best.
5. Thomas S. Kuhn, *The Structure of Scientific Revolutions*, 4th ed. (Chicago: The University of Chicago Press, 2012), Chapters 2–5. Working in Nick Bigelow's CAT (Cooling and Trapping) research group at the University of Rochester was a great window into basic scientific research. I was stretched and challenged in countless ways. Nick was an invaluable mentor and my colleagues in the lab taught me so much about physics, experimental techniques, and perseverance.
6. Steven Shapin, *The Scientific Revolution* (Chicago: The University of Chicago Press, 1996), 1.
7. Lorraine Daston and Peter Galison, *Objectivity* (New York: Zone Books, 2007).
8. The term Whig history has a long and contentious history. Herbert Butterfield's 1931 treatise *The Whig Interpretation of History* (London: Bell and Sons, 1931) popularized the term, but it has been used primarily by historians of science since the 1960s to denigrate historical narratives that seem to lack historical rigor, are unapologetically presentist, and celebrate scientific progress without reservation. For a longer discussion on Whig history as a historiographic category, see Chapter 1.
9. The history of science cannot simply be reduced to a specialization of history. The roots of its disciplinary formation reveal the muddled

interdisciplinarity from which it came. For more discussion on the intradisciplinary dynamics between history and the history of science, see Chapter 1.

10. The debunking of myths like these is something that the history of science has thrived at over the past century. In particular, this short list of myths is based mostly on the contents of two edited volumes. See Ronald L. Numbers, ed. *Galileo Goes to Jail: And Other Myths about Science and Religion* (Cambridge, MA: Harvard University Press, 2009) and its sequel, Ronald L. Numbers and Kostas Kampourakis, eds. *Newton's Apple and other Myths about Science* (Cambridge, MA: Harvard University Press, 2015).

11. Here "science and technology studies" refers to the broad multidisciplinary academic field that goes by several monikers. For example, at Vassar College, where I teach, there is a multidisciplinary program of Science, Technology, and Society (STS), while the large international STS scholarly society is known as the Society for Social Studies of Science (4S).

12. In particular, see Michel Foucault, "Nietzsche, Genealogy, History" in *Language, Counter-Memory, Practice: Selected Essays and Interviews,* ed. Donald F. Bouchard, trans. Donald F. Bouchard and Sherry Simon, 139–164 (Ithaca: Cornell University Press, 1977); Peter Novick, *That Nobel Dream: The "Objectivity Question" and the American Historical Profession* (Cambridge: Cambridge University Press, 1988); Daniel Lord Smail, *On Deep History and the Brain* (Berkeley: University of California Press, 2006); Elizabeth Leane, *Reading Popular Physics: Disciplinary Skirmishes and Textual Strategies* (Farnham, UK: Ashgate Publishing, 2007); Tilman Sauer and Raphael Scholl, eds. *The Philosophy of Historical Case Studies,* Boston Studies in the Philosophy and History of Science (Dordrecht: Springer, 2016); and Agnes Bolinska and Joseph D. Martin, "Negotiating History: Contingency, Canonicity, and Case Studies," *Studies in History and Philosophy of Science, Part A* 80 (2020): 37–46.

13. For a recent account of the importance of these debates see Naomi Oreskes, *Why Trust Science?* (Princeton: Princeton University Press, 2019), 50–54.

14. Katherina Kinzel, "Pluralism In Historiography: A Case Study of Case Studies," in *The Philosophy of Historical Case Studies,* Boston Studies in the Philosophy and History of Science, eds. Tilman Sauer and Raphael Scholl (Dordrecht: Springer, 2016), 123–150.

15. See Chimamanda Ngozi Adichie, "The Danger of a Single Story," filmed July 2009 at TEDGlobal. TED video, 17:32. https://www.ted.

com/talks/chimamanda_ngozi_adichie_the_danger_of_a_single_
story?language=en#t-1114935.

16. Albert Einstein, *Essays in Science*, Dover ed. (New York: Dover Publications, 2009), 111.

17. Bruno Latour, *Science in Action: How to Follow Scientists and Engineers Through Society* (Cambridge, MA: Harvard University Press, 1987), 4.

18. Jop de Vrieze, "Bruno Latour, a Veteran of the 'Science Wars,' Has a New Mission," *Science*Insider, October 10, 2017. https://www.sciencemag.org/news/2017/10/bruno-latour-veteran-science-wars-has-new-mission.

19. De Vrieze, "Bruno Latour."

20. Gerald J. Holton, *Science and Anti-science* (Cambridge, MA: Harvard University Press, 1993), Chapter 6. Holton discusses the long legacy of "anti-science" movements throughout history and how the latest manifestation is really an umbrella term for multiple movements and social forces. For a more up-to-date discussion on the current perceived "war on science," see Shawn Otto, *The War On Science: Who's Waging It, Why It Matters, What We Can Do About It* (Minneapolis, MN: Milkweed Editions, 2016); Naomi Oreskes and Erik M. Conway, *Merchants of Doubt: How a Handful of Scientists Obscured the Truth on Issues from Tobacco Smoke to Global Warming* (New York: Bloomsbury Press, 2010); De Vrieze, "Bruno Latour"; Oreskes, *Why Trust Science?*; and David Michaels, *The Triumph of Doubt: Dark Money and the Science of Deception* (New York: Oxford University Press, 2020).

21. For an analysis of Fauci's comment, see discussion in the Conclusion. Fauci made this comment during an interview. See Anthony Fauci, "Science is Truth," interview by Michael Caputo, *Learning Curve*, HHS.gov Podcasts, June 17, 2020, audio, 36:12. https://www.hhs.gov/podcasts/learning-curve/learning-curve-05-dr-anthony-fauci-science-is-truth.html.

22. Naomi Oreskes refers to this effect as presenting science as a black box, and calls on scientists and scientific institutions to unpack the box so as to demystify it. Naomi Oreskes, "Understanding the Trust (And Distrust) in Science," interview by Ira Flatow, *Science Friday*, NPR, October 11, 2019, audio, 24:47. https://www.sciencefriday.com/segments/naomi-oreskes-why-trust-science/.

23. De Vrieze, "Bruno Latour."

24. De Vrieze, "Bruno Latour."

25. Oreskes, *Why Trust Science?*

26. Oreskes, *Why Trust Science?*, 57.

27. Oreskes, *Why Trust Science?*, 51.

28. This project heeds the advice of renowned historian of science Myles Jackson to realize the political relevance of historical scholarship on the present and to intervene by re-engaging with scientists in critical discourse. See Myles W. Jackson, *The Genealogy of a Gene: Patents, HIV/AIDS, and Race* (Cambridge, MA: MIT Press, 2015), 24.

29. There are stellar examples of scholars who have done this double training, including Thomas Kuhn, Gerald Holton, Silvan S. Schweber, Peter Galison, Naomi Oreskes, and David Kaiser. Although I certainly do not put myself in this rarified company, I can personally attest to the long and arduous character of this path.

30. I am well aware of Latour's use of black boxes, and in particular "Pandora's black box" in Latour, *Science in Action*, 1–17. There is much to value in this earlier analysis, but as Latour rightly points out, this early work can be heavily charged and controversial. In the spirit of forming common ground and letting scientists speak for themselves, I will now grapple with Einstein's juxtaposition and not with Latour's two-faced Janus.

31. For more on scientific imaginaries, see discussion in Chapter 1. See also Sheila Jasanoff and Sang-Hyun Kim, eds., *Dreamscapes of Modernity: Sociotechnical Imaginaries and the Fabrication of Power* (Chicago: The University of Chicago Press, 2015); Sarah Maza, *The Myth of the French Bourgeoisie: An Essay on the Social Imaginary* (Cambridge, MA: Harvard University Press, 2005); and John Tresch, *The Romantic Machine: Utopian Science and Technology after Napoleon* (Chicago: The University of Chicago Press, 2012), 13.

32. For longer discussions on the controversy surrounding Whig history, see Chapter 1 and the Conclusion to this book.

33. Myth-history is a term introduced by physicist Leon Lederman while echoing, and extending, Richard Feynman's use of myth-stories. See Leon Lederman with Dick Teresi, *The God Particle: If the Universe Is the Answer, What Is the Question?* 2nd ed. (New York: Mariner Books, 2006), 412.

34. Lederman with Teresi, *The God Particle*, 412.

35. Chapter 1 traces the origins and explores the rationale behind myth-history as an analytical category.

36. Over the past two decades, there has been a tremendous push to better understand why STEM fields fail to reflect the diversity of our society. Are underrepresented minorities (URMs) dropping out of the pipeline or being pushed? For more on this, see Elaine Seymour and Anne-Barrie Hunter, eds., *Talking about Leaving Revisited: Persistence,*

Relocation, and Loss in Undergraduate STEM Education (Switzerland: Springer, 2019).

37. William H. McNeill, "Mythistory, or Truth, Myth, History, and Historians," *The American Historical Review,* Supplement to Vol. 91, no. 1 (February 1986): 8.

38. For a fascinating discussion on the use of historical case studies in philosophy of science, see Bolinska and Martin, "Negotiating History," 37–46.

39. For effective uses of this approach, see Collins and Pinch's series of Golem books, in particular the first two: Harry Collins and Trevor Pinch, *The Golem: What You Should Know About Science,* 2nd ed. (Cambridge: Cambridge University Press, 2012) and Harry Collins and Trevor Pinch, *The Golem at Large: What You Should Know About Technology* (Cambridge: Cambridge University Press, 2014). More recently, see the wonderful storytelling and analysis of David Kaiser, *Quantum Legacies: Dispatches from an Uncertain World* (Chicago: The University of Chicago Press, 2020).

40. This should not be read as the application of the arbitrary and problematic internal-external boundary that concerned the history of science for so long. For a brief and engaging discussion on this, see Steven Shapin and Simon Schaffer, *Leviathan and the Air-Pump: Hobbes, Boyle, and the Experimental Life,* 2nd ed. (Princeton: Princeton University Press, 2011), xiv–xv.

41. For more on the concept of ripples in social spacetime, see discussion in Chapter 4. See also Harry Collins, *Gravity's Shadow: The Search for Gravitational Waves* (Chicago: The University of Chicago Press, 2004), 13–14 and 84–85. Collins uses a hyphenated form of this concept, "social space-time"; I will be using the unhyphenated form.

42. Michel Foucault, "Nietzsche, Genealogy, History," 139–164.

43. For more on aspirational norms, see Robert K. Merton, "Science and Technology in a Democratic Order," *Journal of Legal and Political Sociology* 1, no. 1 (1942): 115–126 and *The Sociology of Science: Theoretical and Empirical Investigations* (Chicago: University of Chicago Press, 1973). Merton himself recognized that his original norms and the scientific method were idealizations representing how science ought to be practiced and did not accurately reflect the actual practice itself. For more on this dissonance, see Ian I. Mitroff, "Norms and Counter-Norms in a Select Group of the Apollo Moon Scientists: A Case Study of the Ambivalence of Scientists," *American Sociological Review* 39, no. 4 (August 1974): 579–595.

44. Thomas F. Gieryn, "Boundary-Work and the Demarcation of Science from Non-Science: Strains and Interests in Professional Ideologies of Scientists," *American Sociological Review* 48, no. 6 (1983): 781–795.

45. Janet N. Ahn et al., "Even Einstein Struggled: Effects of Learning About Great Scientists' Struggles on High School Students' Motivation to Learn Science," *Journal of Educational Psychology* 108, no. 3 (2016): 314–328.

46. Ahn et al., "Even Einstein Struggled," 314.

47. In the study by Ahn et al., "Even Einstein Struggled," there were two varieties of struggle narrative, but both outperformed the traditional heroic textbook stories. For more on the effects of history on students, see Dennis Farland, "The Effect of Historical, Nonfiction Trade Books on Elementary Students' Perceptions of Scientists," *Journal of Elementary Science Education* 18, no. 2 (Fall 2006): 31–48 and Huang-Yao Hong and Xiaodong Lin-Siegler, "How Learning about Scientists' Struggles Influences Students' Interest and Learning in Physics," *Journal of Educational Psychology* 104 (2012): 469–484.

48. Richard Staley, *Einstein's Generation: The Origins of the Relativity Revolution* (Chicago: The University of Chicago Press, 2009), 420.

49. Mara Beller, *Quantum Dialogue: The Making of a Revolution* (Chicago: The University of Chicago Press, 2001), xiii.

50. Chapter 2 examines a particular case in the history of quantum theory in which an alternate interpretation was filtered out and omitted from discourse and future myth-histories to help reinforce a congealing quantum consensus.

51. Einstein, *Essays in Science*, 111.

52. This is not to say that myth-histories are the sole reason for this mistrust. As Oreskes and Conway, Shawn Otto, David Michaels, and others have unequivocally shown with their research, there are a multitude of factors that are driving this trend. For citations, see note 20.

Chapter 1: Myth-Historical Tensions

1. Errol Morris, *The Ashtray (Or the Man Who Denied Reality)* (Chicago: The University of Chicago Press, 2018), 82.

2. Morris, *The Ashtray,* 9 and xv.

3. Morris, *The Ashtray,* 9 and 13.

4. Morris, *The Ashtray,* xv.

5. I first heard Morris tell a version of this story in a podcast: Barry Lam, "The Ashes of Truth," *Hi-Phi Nation.* S1, episode 9. Podcast

audio, April 18, 2017. https://hiphination.org/complete-season-one-episodes/episode-9-the-ashes-of-truth-april-18-2017/. Since then, Morris has written a memoir detailing his side of the events: Morris, *The Ashtray*, xv–14.

6. Morris, *The Ashtray*, 13. Emphasis in the original. During the calendar year of 1972, Kuhn had taken a leave of absence from Princeton University's Program in the History and Philosophy of Science and was a visiting scholar at the Institute for Advanced Study (IAS). See "Thomas S. Kuhn," Scholars, Institute for Advanced Study (IAS). https://www.ias.edu/scholars/thomas-s-kuhn. According to Morris's memoir, Kuhn was working on his book *Black-Body Theory and the Quantum Discontinuity, 1894–1912*. Morris claims that the alleged confrontation took place at the IAS.

7. Morris, *The Ashtray*, 10–11.

8. Morris, *The Ashtray*, 12–13.

9. Morris, *The Ashtray*, 82.

10. In an interview Morris conducted with Norton Wise, Wise admits to Kuhn's adversarial and borderline threatening demeanor. See Morris, *The Ashtray*, 161–162. It may be that Kuhn was adversarial and even intimidating when engaged in scholarly discourse, but there is just no way to corroborate Morris's story.

11. We will revisit this encounter in the Conclusion. Morris uses his account of the confrontation with Kuhn as a springboard to attack relativism and other currents in contemporary academia. Although Morris makes some interesting points about the slippery slope of relativism, his arguments are shrouded in a single-minded critique of a caricature of Kuhn's ideas and his legacy.

12. Steven Weinberg, *To Explain the World: The Discovery of Modern Science* (London: Penguin UK, 2015), xii–xiii.

13. Weinberg, *To Explain the World*, 29.

14. Weinberg began arguing for, and justifying, his brand of Whiggish history decades ago. See Steven Weinberg, "Sokal's Hoax," *The New York Review of Books*, August 8, 1996. https://www.nybooks.com/articles/1996/08/08/sokals-hoax/. The term Whig history has a long and contentious history. Popularized by Butterfield, in *The Whig Interpretation of History*, it has been used primarily by historians of science since the 1960s to denigrate historical narratives that seem to lack historical rigor, are unapologetically presentist, and celebrate scientific progress without reservation. For more on this, see references in note 40.

15. Butterfield, *The Whig Interpretation of History*, v.

16. Butterfield, *The Whig Interpretation of History*, 31.

17. In leveraging Whig histories, textbooks can be powerful rhetorical tools within scientific communities. They present students with the ideals of scientific practice, the canonical knowledge in a particular field, and they aid in shutting down controversial scientific debates. This process was discussed in Kuhn, *Structure of Scientific Revolutions*, 135–142. More recent discussions can be found in Collins and Pinch, *The Golem*, 153–163; Sharon Traweek, *Beamtimes and Lifetimes: The World of High Energy Physicists* (Cambridge, MA: Harvard University Press, 1988), 74–105; Buhm Soon Park, "In the 'Context of Pedagogy,'" in *Pedagogy and the Practice of Science*, ed. David Kaiser (Cambridge, MA: MIT University Press, 2005), 287–319; David Kaiser, "Turning Physicists Into Quantum Mechanics," *Physics World* (May 2007): 28–33; Suman Seth, *Crafting the Quantum: Arnold Sommerfeld and the Practice of Theory, 1890–1926* (Cambridge, MA: The MIT Press, 2010) especially Chapter 2; and Andrew Warwick, *Masters of Theory: Cambridge and the Rise of Mathematical Physics* (Chicago: University of Chicago Press, 2003).

18. For discussions of "social imaginaries," see Jasanoff and Kim, eds., *Dreamscapes of Modernity*; Maza, *The Myth of the French Bourgeoisie*; and Tresch, *The Romantic Machine*, 13.

19. There is a whole literature dedicated to understanding the history and power of the anecdote and its relationship to historiography. For more on this, see Dorinda Outram, *Four Fools in the Age of Reason: Laughter, Cruelty, and Power in Early Modern Germany* (Charlottesville, VA: University of Virginia Press, 2019); Malina Stefanovska, "Exemplary or Singular? The Anecdote in Historical Narrative," *SubStance* 38, no. 1 (2009): 16–30; Lionel Grossman, "Anecdote and History," *History and Theory* 42 (May 2003): 143–168; and Lee Schweninger, "Clotel and the Historicity of the Anecdote," *MELUS* 24, no. 1 African American Literature (Spring 1999): 21–36.

20. Weinberg, *To Explain the World*, 29. Also see Weinberg, "Sokal's Hoax."

21. Weinberg, "Sokal's Hoax."

22. Weinberg, "Sokal's Hoax."

23. The boundary work mentioned here is in reference to Thomas Gieryn's concept. For further discussion, see Gieryn, "Boundary-Work," 781–795.

24. Steven Shapin, "Why Scientists Shouldn't Write History," review of *To Explain the World*, by Steven Weinberg, *The Wall Street Journal*, February 13, 2015. https://www.wsj.com/articles/book-review-to-explain-the-world-by-steven-weinberg-1423863226.

25. Shapin, *The Scientific Revolution*.

26. Shapin, *The Scientific Revolution*, 1.

27. Emphasis in the original. Weinberg, *To Explain the World*, 146.
28. Collins and Pinch, *The Golem*, 153–163. Apart from textbook histories, Collins and Pinch identify five other types of history of science, including official history, reviewers' history, reflective history, analytic history, and interpretive history. This typology is critical to understanding that not all narrative forms are the same. This chapter argues that myth-history can serve as a powerful underlying category because it is more foundational, it was originally proposed by scientists themselves, and it encompasses much of the dynamics covered in Collins and Pinch's typology.
29. Shapin, "Why Scientists Shouldn't Write History."
30. Allan Franklin and Harry Collins, "Two Kinds of Case Study and a New Agreement," in *The Philosophy of Historical Case Studies*, Boston Studies in the Philosophy and History of Science, eds. Tilman Sauer and Raphael Scholl (Dordrecht: Springer, 2016), 88–115.
31. Akira Kurosawa, *Rashomon* [film]. Producer: Daiei, Japan. Script: T. Matsuama (1950). Kurosawa's masterpiece examines the nature of truth by telling the story of a violent crime from four distinct perspectives. Each version reveals the subjectivity of individual standpoints.
32. For more on the concept of ripples in social spacetime, see discussion in Chapter 4. See also Collins, *Gravity's Shadow*, 13–14 and 84–85.
33. For more on experimenter's regress, see Collins, *Gravity's Shadow*, 126–128.
34. Franklin and Collins, "Two Kinds of Case Study," 89.
35. Franklin and Collins, "Two Kinds of Case Study," 89.
36. Franklin and Collins, "Two Kinds of Case Study," 113.
37. See Chimamanda Ngozi Adichie, "The Danger of a Single Story," filmed July 2009 at TEDGlobal. TED video, 18:34. https://www.ted.com/talks/chimamanda_ngozi_adichie_the_danger_of_a_single_story?language=en#t-1114935.
38. Franklin and Collins, "Two Kinds of Case Study," 89, 91.
39. Franklin and Collins, "Two Kinds of Case Study," 91.
40. There has been much written on this topic over the past several decades. In particular, see David L. Hull, "In Defense of Presentism," *History and Theory* 18, no. 1 (February 1979): 3; Rupert A. Hall, "On Whiggism," *History of Science* 21, no. 1 (1983): 45–59; Edward Harrison, "Whigs, Prigs and Historians of Science," *Nature* 329, no. 6136 (September 17, 1987): 213–214; Adrian Wilson and T.G. Ashplant, "Whig History and Present-Centered History," *The Historical Journal* 31, no. 1 (March 1988): 1–16; Ernst Mayr, "When Is Historiography

Whiggish?," *Journal of the History of Ideas* 51, no. 2 (1990): 301–309; Nick Tosh, "Anachronism and Retrospective Explanation: In Defense of a Present-Centered History of Science," *Studies in History and Philosophy of Science Part A* 34, no. 3 (September 2003): 647–659; Nick Jardine, "Whigs and Stories: Herbert Butterfield and the Historiography of Science," *History of Science* 41, no. 2 (June 1, 2003): 125–140; Oscar Moro-Abadía, "Thinking About 'Presentism' From a Historian's Perspective: Herbert Butterfield and Hélène Metzger," *History of Science*, 47, no. 1 (March 2009): 55–77; Naomi Oreskes, "Why I Am a Presentist," *Science in Context* 26, no. 4 (2013): 595–609; David Alvargonzález, "Is the History of Science Essentially Whiggish?," *History of Science* 51, no. 1 (March 2013): 85–99; Andre Wakefield, "Butterfield's Nightmare: The History of Science as Disney History," *History and Technology* 30, no. 3 (2014): 232–251; Laurent Loison, "Forms of Presentism in the History of Science. Rethinking the Project of Historical Epistemology," *Studies in History and Philosophy of Science* 60, (December 2016): 29–37.

41. For discussion of these historiographic trends, see Novick, *That Nobel Dream*; Anthony Molho and Gordon S. Wood, eds., *Imagined Histories: American Historians Interpret the Past* (Princeton: Princeton University Press, 1998); Smail, *On Deep History and the Brain*; and, most recently, Daniel Steinmetz-Jenkins, "Beyond the Edge of History: Historians' Prohibition on 'Presentism' Crumbles Under the Weight of Events," *The Chronicle of Higher Education*, August 14, 2020. https://www.chron-icle.com/article/beyond-the-end-of-history.

42. John Lewis Gaddis, *The Landscape of History: How Historians Map the Past* (New York: Oxford University Press, 2002), 9–10.

43. Outram, *Four Fools in the Age of Reason,* Kindle ed., Introduction.

44. Gaddis, *The Landscape of History,* 7.

45. Michel Foucault, "Nietzsche, Genealogy, History," 153–155.

46. For further discussion, see Jackson, *The Genealogy of a Gene*, 23–24, and Foucault, "Nietzsche, Genealogy, History," 139–164.

47. From the History of Science Society webpage: "History of the Society," About HSS, History of Science Society (HSS). https://hssonline.org/about/history-of-the-society/.

48. Ken Alder, "The History of Science as Oxymoron: From Scientific Exceptionalism to Episcience," *Isis* 104, no. 1 (March 2013): 92.

49. Alder, "The History of Science as Oxymoron," 92.

50. For a more complete list of affiliations that historians of science hold, see Alder, "The History of Science as Oxymoron," 91.

51. Alder, "The History of Science as Oxymoron," 91.

52. Loraine Daston, "Science Studies and the History of Science," *Critical Inquiry* 35, no. 4 (Summer 2009): 808. For Rosenberg's original argument, see Charles Rosenberg, "Woods or Trees? Ideas and Actors in the History of Science," *Isis* 79, no. 4 (December 1988): 570. For an interesting counterpoint to Daston, also see Peter Dear and Sheila Jasanoff, "Dismantling Boundaries in Science and Technology Studies," *Isis* 101, no. 4 (December 2010): 759–774.

53. Alder, "The History of Science as Oxymoron," 92.

54. Hall, "On Whiggism," 45–59. For a very timely discussion on historians' reimagining of presentism, see Steinmetz-Jenkins, "Beyond the Edge of History."

55. For example: Beller's study of orthodox narratives in *Quantum Dialogue*; Traweek's study of pilgrims' tales of progress in *Beamtimes and Lifetimes*; Staley's analysis of participant histories in *Einstein's Generation*; and Trevor Pinch's study of official histories in "What Does a Proof Do If It Does Not Prove?," in *The Social Production of Scientific Knowledge*, eds. Everett Mendelsohn et al. (Boston: D. Reidel Publishing Co., 1977), 171–215.

56. Harry Collins, *Forms of Life: The Method and Meaning of Sociology* (Cambridge, MA: The MIT Press, 2019), 61.

57. Collins and Pinch, *The Golem*, 153–163.

58. Leslie Bennetts, "A Master of Mythology Is Honored," *The New York Times*, March 1, 1985. http://www.nytimes.com/1985/03/01/books/a-master-of-mythology-is-honored.html.

59. Bennetts, "A Master of Mythology."

60. From here on, I will be referring to McNeill's paper, published two months later, and based on his presidential address. McNeill, "Mythistory, or Truth, Myth, History, and Historians," 1–10.

61. McNeill, "Mythistory, or Truth, Myth, History, and Historians," 1.

62. Novick, *That Nobel Dream*, Chapter 13.

63. McNeill, "Mythistory, or Truth, Myth, History, and Historians," 8.

64. Novick, *That Nobel Dream*, Introduction.

65. Richard Feynman, *QED: The Strange Theory of Light and Matter*, 3rd ed. (Princeton, NJ: Princeton University Press, 2006), 6.

66. Richard Feynman, "Take the World from Another Point of View," Yorkshire Public Television Program (1973), video file 36:41, YouTube, posted by mrtp, May 28, 2015. https://www.youtube.com/watch?v=GNhlNSLQAFE.

67. James Gleick wrote a biography of Feynman in 1993 and titled it *Genius*. Gleick, *Genius: The Life and Science of Richard Feynman* (New York: Vintage Books, 1993).

68. Gleick, *Genius*, 10, 323.
69. One of Feynman's best-known speeches exemplifies this: Richard Feynman, "What Is Science?" Presented at the fifteenth annual meeting of the National Science Teachers Association, 1966, New York City; reprinted as Richard Feynman, "What is Science?," *The Physics Teacher* 7, no. 6 (September 1969): 313–320.
70. Gleick, *Genius*, 39.
71. Cheryl R. Ganz, *The 1933 Chicago World's Fair: A Century of Progress* (Urbana: University of Illinois Press, 2008).
72. Gleick, *Genius*, 41.
73. Waldemar Kaempffert, "Science in 151 Words," *The New York Times*, June 4, 1933, 142.
74. Feynman, "Take the World," 8:27.
75. Richard Feynman, "The Development of the Space-Time View of Quantum Electrodynamics," *Nobel Prize Lecture*, December 11, 1965. https://www.nobelprize.org/prizes/physics/1965/feynman/lecture/.
76. Feynman, "The Development of the Space-Time View."
77. Lederman with Teresi, *The God Particle*, 7. Note: although Lederman wrote this book with science writer Dick Teresi, the entire book is written from Lederman's perspective as a scientist-storyteller. Since the stories and ideas are presented as his, my analysis will reference Lederman and not Teresi.
78. Lederman with Teresi, *The God Particle*, 6.
79. Admittedly, *The God Particle* was one of the most impactful books I read as a fledgling physics major. I still have vivid memories of reading it during the summer of 1994. I was enthralled and inspired by all the anecdotes of brazen hero-scientists forging a path that would eventually lead us to our modern-day understanding of particle physics. Like Feynman before him, Lederman was a master storyteller. He made physics fun and accessible. Using humor and curated anecdote, he made foreign historical contexts disarmingly familiar. Reading it again all these years later, I have a completely new appreciation for his rhetorical craft.
80. Lederman with Teresi, *The God Particle*, 367.
81. Lederman with Teresi, *The God Particle*, 412.
82. Lederman with Teresi, *The God Particle*, 412. In invoking Feynman, Lederman casually morphs Feynman's notion of "myth-story" into the concept of "myth-history."
83. Lederman with Teresi, *The God Particle*, 412.
84. Lederman with Teresi, *The God Particle*, 412.
85. Lederman with Teresi, *The God Particle*, 412.

86. The discussion later in Chapter 1 interrogates myth-history as a chimeric category that has practical use in resolving certain apparent contradictions, such as the tension inherent in Einstein's black box.

87. Is Lederman's collective consciousness a Jungian "collective unconscious" or a Durkheimian "conscience collective"? For further discussion on how to understand Lederman's uses of truth and collective consciousness, see section "An Essential Truth Greater than Truth Itself" later in this chapter.

88. Lederman with Teresi, *The God Particle*, Preface.

89. In addition to Lederman's book, Steven Weinberg published his own fierce defense of the SSC in 1992. Steven Weinberg, *Dreams of a Final Theory* (New York: Pantheon Books, 1992).

90. Daniel J. Kevles, "Good-bye to the SSC: On the Life and Death of the Superconducting Super Collider," *Engineering and Science* 58, no. 2 (1995): 22.

91. In addition to Kevles's analysis in Kevles, "Good-bye to the SSC," 16–25, see Kaiser, *Quantum Legacies,* Chapter 10.

92. Lederman with Teresi, *The God Particle*, 189.

93. Lederman with Teresi, *The God Particle*, 193.

94. Lederman with Teresi, *The God Particle*, 189.

95. Lederman with Teresi, *The God Particle*, 190.

96. There is a long tradition in the philosophy of science and STS studying this demarcation problem. For a thoughtful overview, see Michael Gordin, *The Pseudoscience Wars: Immanuel Velikovsky and the Birth of the Modern Fringe* (Chicago: The University of Chicago Press, 2013), 1–18. Gordin examines the demarcation problem in light of the controversies surrounding Immanuel Velikovsky. His conclusion is that far from being a clearly established line, the demarcation between science and pseudoscience is a highly fungible and historically contingent boundary that scientists are constantly redefining.

97. David Kaiser, *How the Hippies Saved Physics: Science, Counterculture, and the Quantum Revival* (New York: W.W. Norton & Company, 2011). For a follow-up discussion, see Kaiser's newest book, *Quantum Legacies,* Chapter 9.

98. Gieryn, "Boundary-work," 781–795.

99. Lederman with Teresi, *The God Particle*, 192.

100. Lederman with Teresi, *The God Particle*, 192.

101. Lederman with Teresi, *The God Particle*, 192.

102. Thomas S. Kuhn, "The Essential Tension: Tradition and Innovation in Scientific Research?" in *The Essential Tension: Selected Studies in*

Scientific Tradition and Change, ed. Thomas S. Kuhn (Chicago: The University of Chicago Press, 1979), 225–239. Ironically, although individual scientists are portrayed as highly divergent, many myth-histories tend to filter out important social divergences from their narratives to highlight continuity and cohesion in science.

103. Richard Feynman, as quoted in Lederman with Teresi, *The God Particle,* 192–193.

104. Lederman with Teresi, *The God Particle,* 194.

105. Lederman with Teresi, *The God Particle,* 194–197.

106. Lederman with Teresi, *The God Particle,* 197.

107. Lederman with Teresi, *The God Particle,* 197.

108. The term "chimeric" here is being used in its biological sense. Chimerism arises when a single organism is made up of two different genotypes that have been merged, yet remain distinct.

109. McNeill, "Mythistory, or Truth, Myth, History, and Historians," 8.

110. Richard Howells, *The Myth of the Titanic,* Centenary ed. (New York: Palgrave Macmillan, 2012), 72.

111. In particular, see Émile Durkheim, *The Elementary Forms of Religious Life,* trans. Karen E. Fields (New York: Free Press, 1995); Claude Lévi-Strauss. *Myth and Meaning* (New York: Schocken Books, 1978); Mircea Eliade, *The Myth of the Eternal Return: Cosmos and History* (Princeton: Princeton University Press, 1971); C.G. Jung and Robert A. Segal, eds. *Jung on Mythology,* 2nd ed. (Princeton: Princeton University Press, 1998); Bronislaw Malinowski, *Myth in Primitive Psychology* (New York: Read Books Ltd., 2014); Ernest Cassirer, *Language and Myth* (New York: Dover, 1953); Roland Barthes, *Mythologies,* 1952, trans. Anette Lavers (New York: Hill and Wang, 1972); Clifford Geertz, "Deep Play: Notes on the Balinese Cockfight," *Dædalus Journal of the American Academy of Arts and Sciences* 101, no. 1 (Winter 1972): 1–38; Hayden White, *Metahistory,* 40th Anniversary ed. (Johns Hopkins University Press, 2014); Jonathan D. Hill, ed., *Rethinking History and Myth: Indigenous South American Perspectives on the Past* (Urbana: University of Illinois Press, 1988); Joseph Campbell with Bill Moyers, *The Power of Myth* (New York: Anchor Books, 1991); Joanne Rappaport, *The Politics of Memory: Native Historical Interpretation in the Colombian Andes* (Durham: Duke University Press, 1998); Duncan S.A. Bell, "Mythscapes: Memory, Mythology, and National Identity," *The British Journal of Sociology* 54, no. 1 (2003): 63–81. Peter Burke, *History and Social Theory,* 2nd ed. (Cornell: Cornell University Press, 2005), 112–140; and Howells, *The Myth of the Titanic,* 72–106.

112. Burke, *History and Social Theory,* 112–140.

113. Claude Lévi-Strauss, "The Structural Study of Myth," *The Journal of American Folklore* 68, no. 270 (1955): 430–431.
114. Feynman, "Take the World," 8:27.
115. The Hohenberg reference is explained in the Introduction to this book.
116. Hill, *Rethinking History and Myth*, 10. See also Howells, *The Myth of the Titanic*, 86.
117. Rappaport, *The Politics of Memory*, 16.
118. For more on aspirational norms, see Robert K. Merton, "Science and Technology in a Democratic Order," *Journal of Legal and Political Sociology* 1, no. 1 (1942): 115–126; Robert K. Merton, *The Sociology of Science: Theoretical and Empirical Investigations* (Chicago: University of Chicago Press, 1973); and Ian I. Mitroff, "Norms and Counter-Norms in a Select Group of the Apollo Moon Scientists: A Case Study of the Ambivalence of Scientists," *American Sociological Review* 39, no. 4 (August 1974): 579–595.
119. For more on the connection between rational agents and universalism in historical analysis, see Burke, *History and Social Theory*, 116–140.
120. Lederman with Teresi, *The God Particle*, 412.
121. Lederman with Teresi, *The God Particle*, 3.
122. Rappaport, *The Politics of Memory*, 17.
123. Rappaport, *The Politics of Memory*, 17.
124. Lederman with Teresi, *The God Particle*, 412.
125. Burke, *History and Social Theory*, 113.
126. Howells, *The Myth of the Titanic*, 94.
127. Howells, *The Myth of the Titanic*, 88.
128. Jan Assmann, *Religion and Cultural Memory*, trans. Rodney Livingstone (Stanford: Stanford University Press, 2006), 1–30.
129. Assmann, *Religion and Cultural Memory*, 9.
130. Assmann, *Religion and Cultural Memory*, 25.
131. Sheila Jasanoff, "Future Imperfect: Science, Technology and the Imaginations of Modernity," in *Dreamscapes of Modernity: Sociotechnical Imaginaries and the Fabrication of Power*, eds. Sheila Jasanoff and Sang-Hyun Kim (Chicago: The University of Chicago Press, 2015), 5–6.
132. Jasanoff, "Future Imperfect," 5.
133. Benedict Anderson, *Imagined Communities: Reflections on the Origin and Spread of Nationalism*, Reprint ed. (London: Verso, 2016), 4.
134. Anderson, *Imagined Communities*, 6.
135. Jasanoff, "Future Imperfect," 6–7.
136. Anderson, *Imagined Communities*, 5.

137. Jasanoff, "Future Imperfect," 4.

138. Jasanoff, "Future Imperfect," 17.

139. Adichie, "The Danger of a Single Story," 9:34.

140. Jasanoff, "Future Imperfect," 18.

141. Lederman with Teresi, *The God Particle*, 412.

142. For more on Newton's broad interests and writings, see Newton Project, "About the Newton Project." http://www.newtonproject. ox.ac.uk/about-us/newton-project.

143. Roger Chartier, *Inscription and Erasure: Literature and Written Culture from the Eleventh to the Eighteenth Century*, trans. Arthur Goldhammer (Philadelphia: University of Pennsylvania Press, 2007), ix and Chapter 2.

144. Staley, *Einstein's Generation,* 420.

145. Assmann, *Religion and Cultural Memory*, 7–8.

146. In particular, for a captivating discussion on the various uses of history and their respective topologies of power, see Foucault, "Nietzsche, Genealogy, History," 139–164.

147. Collins, *Forms of Life*, 17.

Chapter 2: Myth-Historical Quantum Erasure

1. This chapter is based on a previously published case study. José G. Perillán, "Quantum Narratives and the Power of Rhetorical Omission: An Early History of the Pilot Wave Interpretation of Quantum Theory," *Historical Studies in the Natural Sciences* 48, no. 1 (2018): 24–55.

2. This is an aphorism usually attributed to Richard Feynman, although there does not seem to be evidence for this attribution. There is, however, evidence that Steven Weinberg uttered it in 1987. See Wesley C. Salmon, *Causality and Explanation* (Oxford: Oxford University Press, 1998), 386.

3. The orthodox interpretation of quantum theory arose in the late 1920s and early 1930s. It was not inevitable or unitary, and was eventually named the Copenhagen interpretation by Werner Heisenberg in 1955.

4. The French aristocratic physicist Louis de Broglie developed his alternate pilot wave interpretation of quantum theory in 1927 as part of an extensive research program into the foundations of quantum mechanics during the 1920s. As this chapter explains, the pilot wave interpretation would also later be independently developed by David Bohm and renamed the "hidden variables theory."

5. John S. Bell, "On the Impossible Pilot Wave," *Foundations of Physics* 12, no. 10 (October 1982): 989–999. As reproduced in John S. Bell, *Speakable and Unspeakable in Quantum Mechanics* (Cambridge: Cambridge University Press, 1987), 160.

6. Andrew Whittaker, "John Bell in Belfast: Early Years and Education," in *Quantum Unspeakables,* eds. R.A. Bertlman and A. Zeilinger (Berlin: Springer, 2002), 14–17.

7. Bell, "On the Impossible Pilot Wave," 160.

8. Bell, "On the Impossible Pilot Wave," 160.

9. What is commonly known as Heisenberg's "uncertainty principle" originally appeared as the principle of *ungenauigkeit* (or indeterminacy) in 1927. The shift in terminology came with the English translation to Werner Heisenberg, *The Physical Principles of the Quantum Theory* (Chicago: University of Chicago Press, 1930). For more discussion on terminology, see David C. Cassidy, *Beyond Uncertainty: Heisenberg, Quantum Physics, and the Bomb* (New York: Bellevue Literary Press, 2009), 185.

10. There are many excellent nonspecialist conceptual introductions to quantum theory, its paradoxes, and interpretation debates. See Arthur Fine, *The Shaky Game: Einstein Realism and the Quantum Theory,* 2nd ed. (Chicago: The University of Chicago Press, 1996); Bruce Rosenblum and Fred Kuttner, *Quantum Enigma: Physics Encounters Consciousness,* 2nd ed. (Oxford: Oxford University Press, 2011); Paul Halpern, *Einstein's Dice and Schrodinger's Cat: How Two Great Minds Battled Quantum Randomness to Create a Unified Theory of Physics* (Philadelphia, PA: Basic Books, 2015); and Art Hobson, *Tales of the Quantum: Understanding Physics' Most Fundamental Theory* (New York: Oxford University Press, 2017).

11. This apocryphal statement captures the pragmatism that dominated the physics community after 1930. It is often attributed to Richard Feynman, but more likely originated with David Mermin. See N. David Mermin, "Could Feynman Have Said This?," *Physics Today* 57, no. 5 (2004): 10–11.

12. Beller, *Quantum Dialogue,* 212.

13. Lederman with Teresi, *The God Particle,* 412 and 197.

14. For further discussion on the rise of pragmatism, see Silvan S. Schweber, "The Empiricist Temper Regnant: Theoretical Physics in the United States 1920–1950," *Historical Studies in the Physical and Biological Sciences* 17, no. 1 (1986): 55–98, on 56–58. David Kaiser, "Making Tools Travel: Pedagogy and the Transfer of Skills in Post War Theoretical Physics," in *Pedagogy and the Practice of Science,* ed. David Kaiser (Cambridge, MA: MIT University Press, 2005), 41–74, is an excellent case study of the dissemination of Feynman diagrams as a

pedagogical transfer of a tacit toolkit that emphasized rigor of calculations over interpretational concerns. For further commentary
on the rise of pragmatism in physics, see Olival Freire Junior, *The
Quantum Dissidents: Rebuilding the Foundations of Quantum Mechanics (1950–
1990)* (Berlin: Springer, 2015), Chapter 2.1.

15. Leonard I. Schiff, *Quantum Mechanics* (New York: McGraw-Hill, 1949),
6–8. In 1965, a Berkeley physics professor, Eyvind Wichmann, fondly
recalled the pragmatic and applied nature of Schiff's book: "The
book kept me sufficiently busy to prevent pseudo-philosophical
speculations about the True Meaning of quantum mechanics." This
reference was found in Kaiser, *How the Hippies Saved Physics*, 19.

16. For a fascinating analysis of Schiff's textbook placed into its Cold
War context, see Kaiser, *Quantum Legacies,* Chapter 8.

17. Traweek, *Beamtimes and Lifetimes*, 74–105.

18. Whittaker, "John Bell in Belfast," 14–17. For this reconstruction,
Whittaker relies on an interview conducted by Jeremy Bernstein
with Bell (circa 1990) as well as recollections from Lesley Kerr, a Bell
classmate at Queen's. This episode is also recounted in Kaiser, *How the
Hippies Saved Physics*, 25–26.

19. Bell, "On the Impossible Pilot Wave," 159. Bell refers to his first reading of Max Born, *Natural Philosophy of Cause and Chance* (New York:
Dover Publications, 1964), first published by Oxford University Press
in 1949, 109. This became an important text for Bell's education in
quantum theory, and later a target of his frustrations.

20. Born, *Natural Philosophy of Cause and Chance.*

21. James T. Cushing, *Quantum Mechanics: Historical Contingency and the
Copenhagen Hegemony* (Chicago: The University of Chicago Press, 1994),
xii. Here, Cushing invokes the measurement problem and the classical limit as fundamental and persistent mysteries emerging from
the orthodox interpretation of quantum theory.

22. For a discussion on the myth of a unitary Copenhagen interpret-
ation, see Beller, *Quantum Dialogue*; Don Howard, "Who Invented the
'Copenhagen Interpretation'? A Study in Mythology," *Philosophy of
Science* 71, no. 5 (December 2004): 669–682; Kristian Camilleri,
"Constructing the Myth of the Copenhagen Interpretation,"
Perspectives on Science 17, no. 1 (2009): 26–57; and Kristian Camilleri,
*Heisenberg and the Interpretation of Quantum Mechanics: The Physicist as
Philosopher* (Cambridge: Cambridge University Press, 2009).

23. Mara Beller's study of orthodox narratives in *Quantum Dialogue* can be
situated within the larger study of communal narratives; for
example, Traweek's study of pilgrim's tales of progress in *Beamtimes*

and Lifetimes, Staley's analysis of participant histories in *Einstein's Generation*, and Pinch's study of official histories in "What Does a Proof Do If It Does Not Prove?"

24. Beller, *Quantum Dialogue*, xiii.

25. Beller, *Quantum Dialogue*, 212. This group of physicists refers primarily to those associated with the Copenhagen and Göttingen quantum schools in the 1920s and 1930s. These physicists, who pushed the orthodox interpretation of quantum theory, are also described in John Heilbron, "The Earliest Missionaries of the Copenhagen Spirit," *Revue d'histoire des sciences* 38, no. 3 (1985): 195–230 and in Kaiser, *How the Hippies Saved Physics*, 14. Here Kaiser refers to a "core set" of quantum architects.

26. Staley, *Einstein's Generation*, 420.

27. Lederman with Teresi, *The God Particle*, 412. For a full discussion on how Letterman uses the term myth-history to distinguish it from narratives written by professional historians, see Chapter 1.

28. Born, *Natural Philosophy of Cause and Chance*, 109.

29. A.S. Eddington, "The Future of International Science," *Observatory* 39 (June 1916): 271, quoted in Matthew Stanley, *Practical Mystic: Religion, Science, and A.S. Eddington* (Chicago: The University of Chicago Press, 2007), 88–89.

30. Michael Eckert, "The Emergence of Quantum Schools: Munich, Göttingen and Copenhagen as New Centers of Atomic Theory," *Annalen der Physik* 10 (2001): 151–162.

31. Much has been written about the turbulence of the 1920s within the physics community and the effects of international tensions during the period. For example, see Paul Forman, "Scientific Internationalism and the Weimar Physicists: The Ideology and Its Manipulation in Germany after World War I," *Isis* 64 (1973): 150–180; Daniel Kevles, *The Physicists: The History of a Scientific Community in Modern America* (Cambridge, MA: Harvard University Press, 1995), 185–221; Stanley, *Practical Mystic*, 120–123; Helge Kragh, *Quantum Generations: A History of Physics in the Twentieth Century* (Princeton, NJ: Princeton University Press, 2002), 141–145; Grégoire Wallenborn and Pierre Marage, ed., *The Solvay Councils and the Birth of Modern Physics* (Basel: Birkhäuser, 1999), 113–115; and, most recently Jimena Canales, *The Physicist and the Philosopher: Einstein, Bergson, and the Debate that Changed our Understanding of Time* (Princeton, NJ: Princeton University Press, 2015), 96 and 121–127.

32. For an account of Maurice de Broglie's influence on Louis de Broglie, see Bruce Wheaton, "Atomic Waves in Private Practice," in *Quantum*

Mechanics at the Crossroads, eds. James Evans and Alan Thorndike (Berlin: Springer-Verlag, 2007), Chapter 3. Also see Mary J. Nye, "Aristocratic Culture and the Pursuit of Science: The De Broglies in Modern France," *Isis* 88, no. 3 (September 1997): 397–421 and the dissertation by Adrien Vila-Valls, "Louis de Broglie et la Diffusion de la Mécanique Quantique en France (1925–1960)" (PhD dissertation, Université Claude Bernard, 2012), 52–60.

33. For a longer account of the tension between de Broglie and physicists in Copenhagen, see V.V. Raman and P. Forman, "Why Was It Schrödinger Who Developed de Broglie's Ideas?," *Historical Studies in the Physical Sciences* 1 (1969): 294–295.

34. Within the French physics community of the 1920s, the dominant paradigm was still the established and celebrated tradition of mathematical physics, while the newer emerging tradition of theoretical physics was struggling for institutional traction. For further discussion of isolation and inertia, see Vila-Valls, "Louis de Broglie," 24–46 and 61–65.

35. Born, *Natural Philosophy of Cause and Chance,* 90–91.

36. This was an important insight as it explicitly related particle properties like momentum (p) to wave properties like wavelength (λ) for objects with finite mass.

37. For an analysis of de Broglie's research program during the 1920s, see Guido Bacciagaluppi and Antony Valentini, eds., *Quantum Theory at the Crossroads: Reconsidering the 1927 Solvay Conference* (Cambridge: Cambridge University Press, 2009), Chapter 2.

38. Translation of excerpts from Louis de Broglie's 1924 dissertation (p. 56) as translated and reproduced by Bacciagaluppi and Valentini, *Quantum Theory at the Crossroads,* 43.

39. Raman and Forman, "Why Was It Schrödinger Who Developed de Broglie's Ideas?," 291–314. The authors rigorously analyze why Schrödinger would have been one of the few physicists not prejudiced against de Broglie and his ideas. In addition, they reflect on Schrödinger's previous research and why he was well positioned to build upon de Broglie's ideas.

40. Louis de Broglie, "La Structure Atomique de la Matière et du Rayonnement et la Mécanique Ondulatoire," *Comptes Rendus* 184 (1927): 273–274.

41. Louis de Broglie, "La Mécanique Ondulatoire et la Structure Atomique de la Matière et du Rayonnement," *Journal de Physique et du Radium* 8 (1927): 225–241.

42. There seems to be a parallel between the difficulty de Broglie had in applying his deterministic wave-mechanical formulation of quantum theory based on a hydrodynamic analogy and the failure of early climate models. For more about the problems and failures of early climate modeling, see Spencer Weart, *The Discovery of Global Warming: Revised and Expanded Edition* (Cambridge, MA: Harvard University Press, 2008), Chapter 1, and Kristine Harper, *Weather by the Numbers: The Genesis of Modern Meteorology* (Cambridge, MA: MIT Press, 2008), Chapters 1–2.

43. Beller refers to this period in the development of quantum theory as being in "creative dialogical flux" as she tracks the polyphony of contradictory voices and ideas, many times coming from the same physicist. Beller, *Quantum Dialogue*, 3 and Chapter 8. All of this flux, or epistemological plurality, emerges even before one considers an analysis of alternate deterministic interpretations of quantum theory.

44. See, among others, Max Jammer, *The Conceptual Development of Quantum Mechanics (History of Modern Physics, 1800–1950)* (New York: McGraw-Hill, 1966), 291–293; Darren Belousek, "Einstein's 1927 Unpublished Hidden Variables Theory: Its Background, Context and Significance," *Studies in History and Philosophy of Modern Physics* 21, no. 4 (1996): 437–461; Léon Brillouin, "The New Atomic Mechanics," *Journal de Physique et le Radium* 7 (1926): 135 (reprinted in *Selected Papers on Wave Mechanics*, translated by Winifred Deans, London: Blackie & Son Limited 1928, 52); and O. Klein, "The Atomicity of Electricity as a Quantum Theory Law," *Nature* 118, no. 2971 (1926): 516.

45. H.A. Lorentz had been trying unsuccessfully to end Solvay's exclusionary policy for years, but with the imminent inclusion of Germany in the League of Nations, momentum began to swing his way. For more on Lorentz's exceptional work in guiding the earliest Solvay Councils, see Peter Galison, "Solvay Redivivus," in *Proceedings of the 23rd Solvay Conference on Physics: The Quantum Structure of Space and Time*, eds. David Gross et al. (Singapore: World Scientific Publishing, 2007), 1–18. See also Bacciagaluppi and Valentini, *Quantum Theory at the Crossroads*, 3 and 9. For an interesting analysis of the role of the League of Nations in ending political exclusion of German scientists, see Canales, *The Physicist and the Philosopher*, 96 and 121–127.

46. In most narratives describing the fate of de Broglie's pilot wave theory, there is a critical exchange between Pauli and de Broglie at Solvay in which Pauli describes the physical scenario involving the inelastic scattering of electrons by a rigid rotator. He finds critical

fault in de Broglie's pilot wave picture, which cannot apparently account for experimentally verified scattering outcomes. Their exchange is not included here, because it occurred during the general discussion, well after Born and Heisenberg had made their remarks on closing the debate. Neither Pauli's critique nor de Broglie's response affected the rhetoric of closure discussed here. However, I recognize that Pauli's critique played a role in de Broglie's abandonment of his pilot wave theory. Reference to this is made later in this chapter.

47. Max Born and Werner Heisenberg, "Quantum Mechanics," published in the *Fifth Solvay Council Proceedings* in 1928; reprinted in Bacciagaluppi and Valentini, *Quantum Theory at the Crossroads*, 398.

48. Hendrik A. Lorentz, in "General Discussion of the New Ideas Presented," published in the *Fifth Solvay Council Proceedings* in 1928; reprinted in Bacciagaluppi, and Valentini, *Quantum Theory at the Crossroads*, 432–433. Also see Galison, "Solvay Redivivus", 10–13.

49. As an example, Heisenberg's treatment of the helium spectrum in 1926 was a major contribution that could not have been achieved before the development of the new formalisms of quantum mechanics. See Jagdish Mehra, *The Solvay Conferences on Physics: Aspects of the Development of Physics since 1911* (Boston: D. Reidel Publishing Co., 1975), 134–135.

50. Having said this, we should recall that Beller's dialogical analysis clearly shows individual polyphony dependent on a physicist's particular audience and local context: "Acausality, in the writings of Bohr, Heisenberg, Pauli, and Born, is not a sharply defined concept. Because acausality is to be used as a tool of legitimation, as a sword to wield against opponents, its meaning changes from text to text, from context to context" (Beller, *Quantum Dialogue*, 196; also see Chapters 5 and 8). This idea of individual polyphony is also expressed clearly in an interesting analysis by Alexei Kojevnikov, "Philosophical Rhetoric in Early Quantum Mechanics 1925–27: High Principles, Cultural Values and Professional Anxieties," in *Weimar Culture and Quantum Mechanics: Selected Papers by Paul Forman and Contemporary Perspectives on the Forman Thesis*, eds. Cathryn Carson, Alexei Kojevnikov, and Helmuth Trischler (London: Imperial College Press, 2011), 319–348. Here the tension between individual or local context and the effects of community identity emerges again. We should recognize the power of congealing community consensus without dismissing the individual agency of scientists in their local contexts.

51. For analysis of the emergence of the rhetoric of inevitability and finality leading to a closure of interpretation debates, see Beller, *Quantum Dialogue*, Chapter 9.

52. Oliver Lodge, "Truth or Convenience" *Nature* 119, no. 2994 (1927), 424.

53. From the December 1926 correspondence between Darwin and Bohr in the Archives for the History of Quantum Physics (AHQP), Niels Bohr Library at the American Institute of Physics (AIP), College Park, MD, quoted in Beller, *Quantum Dialogue*, 4.

54. Albert Einstein to Erwin Schrödinger, May 31, 1928, in Karl Przibram, ed., *Letters on Wave Mechanics: Schrödinger, Planck, Einstein, Lorentz* (New York: Philosophical Library, 1967), 31; quoted in Kaiser, *How the Hippies Saved Physics*, 8.

55. Heilbron, "The Earliest Missionaries of the Copenhagen Spirit," 195–230.

56. Heilbron, "The Earliest Missionaries of the Copenhagen Spirit," 196. Over the past thirty years, the extent to which this Copenhagen spirit was uniform has been brought into question. In particular, see note 22 for references.

57. In a letter from Heisenberg to Bohr, July 27, 1928; quoted in Heilbron, "The Earliest Missionaries of the Copenhagen Spirit," 201. For a rich discussion on Heisenberg's evolving quantum philosophy and his fascinating intellectual journey, see Camilleri, *Heisenberg and the Interpretation of Quantum Mechanics*, especially Chapters 4–6. In Camilleri's discussion, one immediately realizes that a characterization of Heisenberg as simply a "disciple" of Bohr and the Copenhagen spirit is far too restrictive. For example, Heisenberg and Bohr used terminology like wave-particle duality and complementarity in very distinct ways, leading to miscommunications and misunderstandings. A careful exploration of their intersecting and evolving views reveals surprising divergence on key conceptual issues. However, these divergences were mostly hidden from the larger physics community. As a result, the outward stance of a resonant Copenhagen spirit was used for real rhetorical effect.

58. Heilbron, "The Earliest Missionaries of the Copenhagen Spirit," 201.

59. Gleick, *Genius*, 39. It seems the teenage Feynman did not coincide with Niels Bohr at the Chicago Fairgrounds during the summer of 1933. For more on Feynman and the Chicago World's Fair, see Chapter 1.

60. In addition to the sources listed, for further discussion on diverging views of the interpretation of quantum theory among early quantum architects, see Olival Freire Jr., "Quantum Controversy and

Marxism," *Historia Scientiarum* 7, no. 2 (1997): 137–152; Olival Freire Jr., "Orthodoxy and Heterodoxy in the Research on the Foundations of Quantum Physics: E.P. Wigner's Case," in *Cognitive Justice in a Global World: Prudent Knowledges for a Decent Life*, ed. Boaventura de Sousa Santos (Langham: Lexington Books, 2007), Chapter 10; Kaiser, *How the Hippies Saved Physics*, Chapter 1; and Anja Skaar Jacobsen, "Léon Rosenfeld's Marxist Defense of Complementarity," *Historical Studies in the Physical and Biological Sciences* 37 Supplement (2007): 3–34.

61. Kenji Ito, "The *Geist* in the Institute: The Production of Quantum Physics in 1930s Japan," in *Pedagogy and the Practice of Science*, ed. David Kaiser, 151–183.

62. Beller, *Quantum Dialogue*, Chapter 1.

63. Howard, "Who Invented the 'Copenhagen Interpretation'? A Study in Mythology," 675.

64. Beller, *Quantum Dialogue*, xiii.

65. Cushing, *Quantum Mechanics*.

66. Silvan Schweber, *QED and the Men Who Made It* (Princeton: Princeton University Press, 1994), especially Chapters 1–3.

67. P.A.M. Dirac, "Quantum Mechanics of Many-Electron Systems," *Proceedings of the Royal Society London* A123 (1929): 714–733, quoted in Schweber, *QED and the Men Who Made It*, xxii.

68. Max Jammer terms this an "almost unchallenged monocracy of the Copenhagen school"; Jammer, *The Philosophy of Quantum Mechanics: The Interpretations of Quantum Mechanics in Historical Perspective* (New York: John Wiley & Sons, 1974), 247–251, quoted in Freire Jr., "Orthodoxy and Heterodoxy," 205. We must be careful here to emphasize the public dimension of this discourse. Beller shows that in private correspondence and before select audiences, these physicists tended to showcase polyphonic voices on interpretational matters that at times were self-contradictory (Beller, *Quantum Dialogue*, Chapter 8).

69. Jammer, *The Philosophy of Quantum Mechanics*, 247–248.

70. Arnold Sommerfeld, *Atombau und Spektrallinien, Wellenmechanischer Ergonzungsband* (Braunschweig: Vieweg, 1929), v, 299–301, quoted in Heilbron, "The Earliest Missionaries of the Copenhagen Spirit," 206.

71. Seth, *Crafting the Quantum*, 241 and 247–249.

72. Heisenberg, *The Physical Principles of the Quantum Theory*, 66.

73. Kragh, *Quantum Generations*, 211.

74. Vila-Valls, "Louis de Broglie et la Diffusion de la Mécanique Quantique," 65–66.

75. See Nye, "Aristocratic Culture and the Pursuit of Science," 397–421. Also see Vila-Valls, "Louis de Broglie et la Diffusion de la Mécanique

Quantique," 192–212, in which the author explores de Broglie's effects on the diffusion of quantum theory within the context of French science.

76. An analysis of de Broglie's scientific reasons for abandonment can be found in Bacciagaluppi and Valentini, *Quantum Theory at the Crossroads*, 224–241. De Broglie had essentially taken the pilot wave as far as he could analytically and he understood that the other formulations seemed much more powerful. In particular, while de Broglie had to deal with measurement apparatus within his theory, Heisenberg et al. assumed an instantaneous collapse of the wave function, thereby treating measurement apparatus as external classical objects outside of quantum theory. This simplified calculations greatly. De Broglie realized there were fundamental difficulties (like measurement) with his provisional idea of the pilot wave.

77. Louis de Broglie, *Matter and Light—The New Physics*, trans. W.H. Johnston (New York: Norton and Co., first published in 1939), 189.

78. Louis de Broglie, *An Introduction to the Study of Wave Mechanics*, trans. H.T. Flint (London: Methuen & Co. Ltd., 1930), 7.

79. De Broglie, *Matter and Light*, 190.

80. One should note that the 1935 Einstein-Podolsky-Rosen paper introducing the EPR paradox and Schrödinger's cat paradox proposed that same year are certainly examples of prominent physicists attempting to keep quantum interpretational issues open in order to counteract a rising tide of pragmatism. However, these attempts were far from the norm, and the paradoxes they created were not seriously grappled with by mainstream physicists until decades later. For more on these paradoxes and their context, see Jammer, *The Philosophy of Quantum Mechanics*, 217–247; Arthur Fine, *The Shaky Game* (Chicago: The University of Chicago Press, 1996), especially Chapters 3 and 5; and Kaiser, *Quantum Legacies*, Chapter 2.

81. For more on pragmatism, World War II, and the Cold War, see Kevles, *The Physicists*, Chapters 20, 21; Schweber, "The Empiricist Temper Regnant: Theoretical Physics in the United States 1920–1950"; Kai Bird and Martin J. Sherwin, *American Prometheus: The Triumph and Tragedy of J. Robert Oppenheimer* (New York: Knopf Publishing, 2005), Part 3; and Jessica Wang, *American Science in an Age of Anxiety: Scientists, Anticommunism, and the Cold War* (Chapel Hill: University of North Carolina Press, 1999), Chapter 1.

82. J. Von Neumann, *Mathematische Grundlagen der Quantenmechanik*. First published in German in 1932 and then translated as *Mathematical Foundations of Quantum Mechanics* (Princeton: Princeton University Press, 1955).

83. Von Neumann, *Mathematical Foundations of Quantum Mechanics,* 324–325.

84. Pinch, "What Does a Proof Do If It Does Not Prove?," 208.

85. Max Born, "Hilbert und die Physik," *Die Naturwissenschaften* 10 (1922): 88–93, translated and quoted in Michael Stöltzner, "Opportunistic Axiomatics—von Neumann on the Methodology of Mathematical Physics," in *John von Neumann and the Foundations of Quantum Physics,* eds. M. Rédei and M. Stöltzner (Dordrecht: Kluwer Academic Publishers, 2001), 39.

86. Dennis Dieks, "Von Neumann's Impossibility Proof: Mathematics in the Service of Rhetorics," *Studies in History and Philosophy of Science Part B* 60, (November 2017): 136–148. In his study of the legacy of von Neumann's impossibility proof, Dieks convincingly problematizes the traditional and simplistic narrative of a unitary Copenhagen hegemony employing von Neumann's impossibility proof as a way of preserving its authority and marginalizing all hidden variables theories. He cites historical evidence of repeated misrepresentations of von Neumann's proof to rhetorically attack the orthodox Copenhagen interpretation. However, all of these rhetorical attacks came about well after 1960. Before then, there is evidence of physicists using the impossibility proof rhetorically to shut down discourse around hidden variables theories. I agree with Dieks that it wasn't a coordinated marginalization from a quantum hegemony, but it did happen. For particular examples of this rhetoric from Pauli, Rosenfeld, and others, see Pinch, "What Does a Proof Do If It Does Not Prove?," 171–209; Beller, *Quantum Dialogue,* 213–214; and F. David Peat, *Infinite Potential: The Life and Times of David Bohm* (New York: Basic Books, 1997), 129–133.

87. Born, *Natural Philosophy of Cause and Chance,* 109.

88. Bell, "On the Impossible Pilot Wave," 160.

89. Bell, "On the Impossible Pilot Wave," 160. Emphasis added.

90. Bell, "On the Impossible Pilot Wave," 160.

91. David Bohm, "A Suggested Interpretation of Quantum Theory in Terms of 'Hidden' Variables. I," *Physical Review* 85, no. 2 (January 1952): 166–179, and "A Suggested Interpretation of Quantum Theory in Terms of 'Hidden' Variables. II," *Physical Review* 85, no. 2 (January 1952): 180–193.

92. Bohm, "A Suggested Interpretation of Quantum Theory in Terms of 'Hidden' Variables. I," 170.

93. Bohm's contempt charges were based on his refusal to answer questions at hearings by the U.S. House of Representatives Committee

on Un-American Activities (HCUA) in the spring of 1949. This infamous committee has come to be known by the acronym HUAC; I will use the more popular acronym. Bohm was not a direct target of the HUAC investigations, but he became collateral damage for not cooperating with the committee. See full transcript of the hearings: Committee on Un-American Activities—House of Representatives, U.S. Congress. "Hearings Regarding Communist Infiltration of Radiation Laboratory and Atomic Bomb Project at the University of California, Berkeley, Calif. Vols. 2–3." Printed at Washington, DC: United States Government Printing Office, 1951. Also see Russell Olwell, "Physical Isolation and Marginalization in Physics: David Bohm's Cold War Exile," *Isis* 90, no. 4 (December, 1999), 745–746.

94. David Joseph Bohm, FBI FOIA File #100-207045, 13, 24.

95. For full biographies of Bohm, see Peat, *Infinite Potential* and Olival Freire Junior, *David Bohm: A Life Dedicated to Understanding the Quantum World* (Switzerland: Springer, 2019). For shorter snapshots and biographical sketches, see Olival Freire Junior, *The Quantum Dissidents*, Chapter 2; Olwell, "Physical Isolation and Marginalization in Physics"; Freire Jr., "Science and Exile: David Bohm, the Hot Times of the Cold War, and His Struggle for a New Interpretation of Quantum Mechanics," *Historical Studies in the Physical and Biological Sciences* 36, no. 1 (September 2005): 1–34; Alexei Kojevnikov, "David Bohm and Collective Movements," *Historical Studies in the Physical and Biological Sciences* 33, no. 1 (2002): 161–192; Christian Forstner, "The Early History of David Bohm's Quantum Mechanics through the Perspective of Ludwik Fleck's Thought Collectives," *Minerva* 46, no. 2 (2008): 215–229; and José G. Perillán, "A Reexamination of Early Debates on the Interpretation of Quantum Theory: Louis de Broglie to David Bohm" (PhD dissertation, University of Rochester, 2011), Chapter 6.

96. Silvan S. Schweber, "Shelter Island, Pocono, and Oldstone: The Emergence of American Quantum Electrodynamics after World War II," *Osiris* 2, no. 1 (1986): 265–302.

97. Schweber, "Shelter Island, Pocono, and Oldstone," 275.

98. The Nobel Prize was awarded to the three physicists for "their fundamental work in quantum electrodynamics, with deep-ploughing consequences for the physics of elementary particles." See "The Nobel Prize in Physics 1965," Nobel Prizes and Laureates, NobelPrize. org. https://www.nobelprize.org/prizes/physics/1965/summary/.

99. Olwell, "Physical Isolation and Marginalization in Physics," 746–747.

100. Ironically, Bohm and Leonard Schiff had both studied with J.R. Oppenheimer, but their approach to teaching quantum theory could not be more different. For a discussion, see Kaiser, *Quantum Legacies*, Chapter 8. Kaiser compares Bohm's textbook to Schiff's *Quantum Mechanics*, carefully situating both in their Cold War contexts.

101. Bohm, *Quantum Theory* (New York: Prentice-Hall, 1951).

102. Bohm's textbook includes a 172-page discussion on conceptual and philosophical foundations, and only then turns to more formal mathematical formulations and applications.

103. For a more detailed analysis of Bohm's textbook and these "latent seeds," see Perillán, "A Reexamination," 318–327.

104. Rosenblum and Kuttner, *Quantum Enigma,* 3.

105. Peat, *Infinite Potential,* 102–109, and Olwell, "Physical Isolation and Marginalization in Physics," 747–748.

106. Olwell, "Physical Isolation and Marginalization in Physics," 750.

107. Freire Junior, *The Quantum Dissidents*, 33, and Peat, *Infinite Potential,* 116–117.

108. Much has been written on Bohm's hidden variables papers and their reception. See the extensive work by Freire Junior, especially *The Quantum Dissidents*, Chapters 2.3–2.4. Also see Peat, *Infinite Potential,* 113 and 125, and Kojevnikov, "David Bohm and Collective Movement," 161–192.

109. Letter from Léon Rosenfeld to David Bohm, dated May 30, 1952. Folder C58, David Bohm Papers, The Library, Birkbeck, University of London, UK.

110. Peat, *Infinite Potential,* 133.

111. Freire Junior, *The Quantum Dissidents*, 167 and Chapter 7.

112. As noted earlier, de Broglie had little difficulty changing his mind. Inspired by Bohm's accomplishments in the early 1950s, he eventually resumed research on the pilot wave theory and once again became one of its most vocal champions.

113. The reader should be cautioned that this was not a linear progression. The ebbs and flows of quantum interpretational debates is a fascinating topic. For more on this, see Kaiser, *Quantum Legacies*, and Freire Junior, *The Quantum Dissidents*.

114. Lederman with Teresi, *The God Particle*, 197.

115. John W.M. Bush, "The New Wave of Pilot-Wave Theory," *Physics Today* 68, no. 8 (August 2015): 47–53.

Chapter 3: Myth-Historical CRISPR Edits

1. The power of storytelling is central to the epic HBO series *Game of Thrones*. This statement comes from one of the last scenes from the

series finale. See Andrew R. Chow, "The Significance of Brienne's Tribute to Jaime in the Game of Thrones Finale," *Time Magazine*, May 20, 2019. https://time.com/5591842/game-of-thrones-brienne-finale/.

2. Eric S. Lander, "The Heroes of CRISPR," *Cell* 164, no. 1–2 (January 14, 2016): 18–28.

3. Four days after Lander's piece appeared in *Cell*, Nathanial Comfort wrote a blog post titled "A Whig History of CRISPR." He framed Lander's history as a clear example of presentist Whig history. Nathanial Comfort, "A Whig History of CRISPR," *Genotopia: Here Lies Truth*, published online January 18, 2016. https://genotopia.scienceblog.com/573/a-whig-history-of-crispr/.

4. Michael Eisen, "The Villain of CRISPR," michaeleisen.org [blog]; published online January 25, 2016. http://www.michaeleisen.org/blog/?p=1825.

5. Eisen, "The Villain of CRISPR."

6. For discussion of Lederman, scientist-storytellers, and their myth-histories, see Chapter 1.

7. Comfort, "A Whig History of CRISPR."

8. Comfort, "A Whig History of CRISPR."

9. As someone trained in physics and history, I pretend no expertise on the underlying biochemical mechanisms of CRISPR and its associated enzymes. In researching this case study, I found many resources that were clear and informative about the biochemical foundations necessary to navigate the controversy. In particular, regardless of the myth-historical nature of the two narratives I'm focusing on here, they both effectively cover the basic science of CRISPR. See Lander, "The Heroes of CRISPR" and Jennifer A. Doudna and Samuel H. Sternberg, *A Crack in Creation: Gene Editing and the Unthinkable Power to Control Evolution* (Boston: Houghton Mifflin Harcourt, 2017). In addition, the following sources were helpful: James Kozubek, *Modern Prometheus: Editing the Human Genome with CRISPR-Cas9* (Cambridge: Cambridge University Press, 2018), and R.M. Gupta and K. Musunuru, "Expanding the Genetic Editing Tool Kit: ZFNs, TALENs, and CRISPR-Cas9," *Journal of Clinical Investigation* 124, no. 10 (October 2014): 4154–4161.

10. Dana Carroll as quoted in John Travis, "Breakthrough of the Year: CRISPR Makes the Cut," *ScienceMag*, December 17, 2015. http://www.sciencemag.org/news/2015/12/and-science-s-2015-breakthrough-year.

11. For discussion of applications of CRISPR-Cas9 in plants and livestock, see Doudna and Sternberg, *A Crack in Creation*, Chapters 4–5.

12. For applications in cancer immunotherapy, see Hong-yan Wu and Chun-yu Cao, "The Application of CRISPR-Cas9 Genome Editing Tool in Cancer Immunotherapy," *Briefings in Functional Genomics* 18, no. 2 (March 22, 2019): 129–132.

13. This list comes from an online article that reviews recent applications of CRISPR technology; Clara Rodríguez Fernández, "10 Unusual Applications of CRISPR Gene Editing," *Labiotech,* March 4, 2019. https://www.labiotech.eu/tops/crispr-applications-gene-editing/.

14. Hank Greely, quoted in *The New Yorker*; Michael Specter, "The Gene Hackers: A Powerful New Technology Enables Us to Manipulate Our DNA More Easily than Ever Before," *The New Yorker*, November 8, 2015. https://www.newyorker.com/magazine/2015/11/16/the-gene-hackers?irgwc=1&source=affiliate_impactpmx_12f6tote_desktop_VigLink&mbid=affiliate_impactpmx_12f6tote_desktop_VigLink.

15. For example, the "DIY Bacterial Gene Engineering CRISPR Kit" is being sold for $169 at *The Odin* website. http://www.the-odin.com/diy-crispr-kit/.

16. Lesley Goldberg, "Jennifer Lopez Sets Futuristic Bio-Terror Drama at NBC (Exclusive)," *The Hollywood Reporter,* October 18, 2016. https://www.hollywoodreporter.com/live-feed/jennifer-lopez-sets-futuristic-bio-939509.

17. Jon Cohen, "How the Battle Lines over CRISPR Were Drawn," *ScienceMag* 2, February 15, 2017. http://www.sciencemag.org/news/2017/02/how-battle-lines-over-crispr-were-drawn.

18. Lander, "The Heroes of CRISPR," 18.

19. Lander, "The Heroes of CRISPR," 18.

20. Lander, "The Heroes of CRISPR," 18.

21. Lander, "The Heroes of CRISPR," 18.

22. Lander, "The Heroes of CRISPR," 20.

23. Lander, "The Heroes of CRISPR," 20.

24. Lander, "The Heroes of CRISPR," 20.

25. Lander, "The Heroes of CRISPR," 26.

26. The Royal Swedish Academy of Sciences, "The Nobel Prize in Chemistry 2020—Genetic Scissors: A Tool for Rewriting the Code of Life," press release, October 7, 2020. https://www.kva.se/en/pressrum/pressmeddelanden/nobelpriset-i-kemi-2020.

27. Tracy Vance, "'Heroes of CRISPR' Disputed," *The Scientist: Exploring Life, Inspiring Innovation,* January 19, 2016. https://www.the-scientist.com/news-opinion/heroes-of-crispr-disputed-34188.

28. One might argue that, due to his proximity, Lander was legally restricted from mentioning the ongoing patent litigation. But that

restriction itself might be reason enough to disqualify him as the CRISPR historian of record. At a minimum, Lander and the editors at *Cell* should have made his interests and those restrictions explicit.

29. Jon Cohen, "With Prestigious Prize, an Overshadowed CRISPR Researcher Wins the Spotlight," *ScienceMag* 6, June 4, 2018. https://www.sciencemag.org/news/2018/06/prestigious-prize-overshadowed-crispr-researcher-wins-spotlight.

30. Doudna and Charpentier's landmark article appeared online in *Science* on June 28, 2012: Martin Jinek, Krzysztof Chylinski, Ines Fonfara, Michael Hauer, Jennifer A. Doudna, and Emmanuelle Charpentier, "A Programmable Dual-RNA–Guided DNA Endonuclease in Adaptive Bacterial Immunity," *Science* 337, no. 6096 (2012): 816–821. Note that there is also a priority dispute associated with this article that will be examined in the following section.

31. According to an iRunway market study titled "CRISPR: Global Patent Landscape," the global market for 2017 was $3.19 billion and was expected to balloon to $6.28 billion by 2022. Adam Houldsworth, "Who Owns the Most CRISPR Patents Worldwide? Surprisingly, It's Agrochemical Giant DowDuPont," *Genetic Literacy Project: Science Not Ideology*, February 16, 2018. https://geneticliteracyproject.org/2018/02/16/owns-crispr-patents-worldwide-surprisingly-agrochemical-giant-dowdupont/.

32. Heidi Ledford, "Titanic Clash over CRISPR Patents Turns Ugly," *Nature News* 537, no. 7621 (2016): 460. https://www.nature.com/news/titanic-clash-over-crispr-patents-turns-ugly-1.20631.

33. Ledford, "Titanic Clash over CRISPR Patents Turns Ugly," 460.

34. United States Patent and Trademark Office (USPTO), "General Information Concerning Patents," October 2015. https://www.uspto.gov/patents-getting-started/general-information-concerning-patents.

35. All these papers appeared in various journals on January 29, 2013. Kozubek, *Modern Prometheus,* 22–24; Doudna and Sternberg, *A Crack in Creation*, 96.

36. In patent applications of this nature, institutions as well as individuals can be named. In the case of this particular application, the Regents of the University of California are named in addition to Doudna and Charpentier.

37. The patent application number in question is 61/652,086; see Jacob S. Sherkow, "Patent Protection for CRISPR: An ELSI Review," *Journal of Law and the Biosciences* 4, no. 3 (2017): 565–576.

38. Jacob S. Sherkow, "Law, History and Lessons in the CRISPR Patent Conflict," *Nature Biotechnology* 33, no. 3 (2015): 256–257.

39. U.S. Patent No. 8,697,359. See Sherkow, "Law, History and Lessons in the CRISPR Patent Conflict."

40. Jon Cohen, "Round One of CRISPR Patent Legal Battle Goes to the Broad Institute," *ScienceMag* 2, February 15, 2017. http://www.sciencemag.org/news/2017/02/round-one-crispr-patent-legal-battle-goes-broad-institute.

41. Jon Cohen, "Federal Appeals Court Hears CRISPR Patent Dispute," *ScienceMag*, 4, April 30, 2018. http://www.sciencemag.org/news/2018/04/federal-appeals-court-hears-crispr-patent-dispute.

42. Jacob S. Sherkow, "The CRISPR Patent Decision Didn't Get the Science Right. That Doesn't Mean It Was Wrong," *STAT*, September 11, 2018. https://www.statnews.com/2018/09/11/crispr-patent-decision-science/.

43. Sherkow, "The CRISPR Patent Decision."

44. For a fascinating discussion of the complex terrain of biotech patents and their relationship with biocapitalism, see Myles Jackson's genealogy of the CCR5 gene: Jackson, *The Genealogy of a Gene.*

45. Kerry Grens, "UC Berkeley Team to Be Awarded CRISPR Patent," *The Scientist: Exploring Life, Inspiring Innovation,* February 11, 2019. https://www.the-scientist.com/news-opinion/uc-berkeley-team-to-be-awarded-crispr-patent-65453.

46. Public Affairs, UC Berkeley, "U.S. Patent Office Indicates It Will Issue Third CRISPR Patent to UC," *Berkeley News*, February 8, 2019. https://news.berkeley.edu/2019/02/08/u-s-patent-office-indicates-it-will-issue-third-crispr-patent-to-uc/.

47. Houldsworth, "Who Owns the Most CRISPR Patents Worldwide?"; Jacob S. Sherkow, "The CRISPR Patent Landscape: Past, Present, and Future," *The CRISPR Journal* 1, no. 1 (February 1, 2018): 5–9; and Heidi Ledford, "How the US CRISPR Patent Probe Will Play Out," *Nature* 531, no. 7593 (March 2016): 149. https://www.nature.com/news/how-the-us-crispr-patent-probe-will-play-out-1.19519.

48. Timothé Cynober, "CRISPR: One Patent to Rule Them All," *Labiotech,* February 11, 2019. https://www.labiotech.eu/in-depth/crispr-patent-dispute-licensing/.

49. Kelly Servick, "Broad Institute Takes a Hit in European CRISPR Patent Struggle," *ScienceMag* 1, January 18, 2018. https://www.sciencemag.org/news/2018/01/broad-institute-takes-hit-european-crispr-patent-struggle; and Jef Akst, "EPO Revokes Broad's CRISPR Patent," *The Scientist: Exploring Life, Inspiring Innovation,* January 17, 2018. https://www.the-scientist.com/the-nutshell/epo-revokes-broads-crispr-patent-30400.

50. See interactive image that illustrates "Dividing the Pie" in Cohen, "How the Battle Lines over CRISPR Were Drawn." Also see Jorge L. Contreras and Jacob S. Sherkow, "CRISPR, Surrogate Licensing, and Scientific Discovery," *Science* 355, no. 6326 (2017): 698–700, and Aggie Mika, "Flux and Uncertainty in the CRISPR Patent Landscape," *The Scientist: Exploring Life, Inspiring Innovation*, October 2017. https://www.the-scientist.com/bio-business/flux-and-uncertainty-in-the-crispr-patent-landscape-30228.

51. Rajendra K. Bera, "The Story of the Cohen-Boyer Patents," *Current Science* 96, no. 6 (March 25, 2009): 760–763. See also Anatole Krattiger et al., *Intellectual Property Management in Health and Agricultural Innovation: A Handbook of Best Practices, Vol. 1* (Oxford, UK: MIHR and Davis, CA: PIPRA, 2007), 1797–1807. Ironically, this is a myth-history itself. In the lead-up to the discovery of rDNA, Janet E. Mertz was a grad student working in Paul Berg's laboratory on rDNA. They published papers in 1972 that showed an easy method to splice together DNA molecules with high efficiency, generating rDNA. When Boyer and Cohen first applied for their rDNA patent in 1974, there was no mention of the work done by Mertz and others in the Berg lab. This controversy is not part of the "gold standard" myth-history. See Kozubek, *Modern Prometheus*, 30.

52. Cohen, "How the Battle Lines over CRISPR Were Drawn."

53. This appears on the National Library of Medicine website. After a quick overview of genome editing and CRISPR-Cas9 in particular, the authors list several resources for further reading, among them Lander's myth-history. National Institutes of Health (NIH), "What Are Genome Editing and CRISPR-Cas9?" https://ghr.nlm.nih.gov/primer/genomicresearch/genomeediting.

54. Lander, "The Heroes of CRISPR," 24.

55. Lander, "The Heroes of CRISPR," 24.

56. Lander, "The Heroes of CRISPR," 24.

57. Lander, "The Heroes of CRISPR," 24.

58. Lander, "The Heroes of CRISPR," 24.

59. Lander, "The Heroes of CRISPR," 24.

60. Lander, "The Heroes of CRISPR," 25.

61. All of these papers appeared in various journals on January 29, 2013. Kozubek, *Modern Prometheus*, 22–24 and Doudna and Sternberg, *A Crack in Creation*, 96.

62. See timelines in both Lander, "The Heroes of CRISPR," 24, and Cohen, "With Prestigious Prize, an Overshadowed CRISPR Researcher Wins the Spotlight."

63. Doudna and Sternberg, *A Crack in Creation,* 85.

64. Alan G. Gross, "Do Disputes over Priority Tell Us Anything about Science?," *Science in Context* 11, no. 2 (Summer 1998): 161–179.

65. Much work has been done to understand and study scientific priority disputes. See Robert K. Merton, "Priorities in Scientific Discoveries: A Chapter in the Sociology of Science," *American Sociological Review* 22 (1957): 635–659; Kuhn, *The Structure of Scientific Revolutions*; Steve Woolgar, "Writing an Intellectual History of Scientific Development: The Use of Discovery Accounts," *Social Studies of Science* 6 (1976): 395–422; Harry M. Collins, "The Place of the 'Core-set' in Modern Science: Social Contingency with Methodological Property in Science," *History of Science* 19, no. 1 (1981): 6–19; Gross, "Do Disputes over Priority?"; and, as it relates to biotechnical patents, Jackson, *The Genealogy of a Gene,* especially Chapters 2–5.

66. Genentech, press release dated January 21, 2009, "Genentech Announces Vice President Appointment in Research." https://www.fiercebiotech.com/biotech/genentech-announces-vice-president-appointment-research.

67. Doudna is listed on the *ScienceMag.org* (AAAS) webpage "Editors and Advisory Boards" under the heading "Board of Reviewing Editors." https://www.sciencemag.org/about/editors-and-editorial-boards.

68. Information for prospective authors is provided on the *ScienceMag.org* (AAAS) webpage "Science: Information for Authors." https://www.sciencemag.org/authors/science-information-authors.

69. ScienceMag.org (AAAS), "Science: Information for Authors."

70. ScienceMag.org (AAAS), "Science: Information for Authors."

71. ScienceMag.org (AAAS), "Science: Information for Authors."

72. ScienceMag.org (AAAS), "Science: Information for Authors."

73. Michael Strevens, "The Role of the Priority Rule in Science," *The Journal of Philosophy* 100, 3 (2003): 55–79.

74. Doudna and Sternberg explicitly make this point in their book *A Crack in Creation,* 91.

75. Kozubek, *Modern Prometheus,* 21.

76. The annual Breakthrough Prizes are awarded to leading thinkers in fundamental physics, life sciences, and mathematics. The foundation is based in Silicon Valley and the prizes are sponsored by leading philanthropists, including Yuri and Julia Milner, Sergey Brin, Anne Wojcicki, Mark Zuckerberg, Priscilla Chan, and Pony Ma. For more information, see the Breakthrough Prize Foundation press release, "Recipients of the 2015 Breakthrough Prizes in Fundamental Physics

and Life Sciences Announced," *Breakthroughprize.org*, November 9, 2014. https://breakthroughprize.org/News/21.

77. Cohen, "With Prestigious Prize, an Overshadowed CRISPR Researcher Wins the Spotlight."

78. Kavli Prize press release, "2018 Kavli Prize in Nanoscience: A Conversation with Jennifer Doudna, Emmanuelle Charpentier and Virginijus Šikšnys," *Nanoscience*, November 9, 2018. http://kavliprize.org/events-and-features/2018-kavli-prize-nanoscience-conversation-jennifer-doudna-emmanuelle-charpentier.

79. Lander, "The Heroes of CRISPR," 26.

80. Lander, "The Heroes of CRISPR," 26.

81. David Beede et al., "Women in STEM: A Gender Gap to Innovation," U.S. Department of Commerce, Economics and Statistics Administration, ESA Issue Brief #04–11 August 2011, 1.

82. M. Allison Kanny, Linda J. Sax, and Tiffani A. Riggers-Piehl, "Investigating Forty Years of STEM Research: How Explanations for the Gender Gap Have Evolved over Time," *Journal of Women and Minorities in Science and Engineering* 20, 2 (2014): 142.

83. Kanny, Sax, and Riggers-Piehl, "Investigating Forty Years of STEM Research," 138–139.

84. Beede et al., "Women in STEM," 8.

85. Kanny, Sax, and Riggers-Piehl, "Investigating Forty Years of STEM Research," 128.

86. Doudna and Sternberg, *A Crack in Creation*, xx.

87. Doudna and Sternberg, *A Crack in Creation*, xx.

88. Sabin Russell, "Cracking the Code: Jennifer Doudna and Her Amazing Molecular Scissors," Cal Alumni Association, UC Berkeley, Winter 2014. https://alumni.berkeley.edu/california-magazine/winter-2014-gender-assumptions/cracking-code-jennifer-doudna-and-her-amazing.

89. Doudna and Sternberg, *A Crack in Creation*, 34.

90. Einstein, *Essays in Science*, 111.

91. Doudna and Sternberg, *A Crack in Creation*, 10.

92. As discussed in James D. Watson, *The Annotated and Illustrated Double Helix* (New York: Simon & Schuster, 2012), Appendix 4. In this appendix, the editors unpack the controversy surrounding the publication of *The Double Helix* in 1968. The book was listed seventh on the Modern Library's list of twentieth-century works of nonfiction and was also included in the Library of Congress list of "Books that Shaped America."

93. Watson, *The Annotated and Illustrated Double Helix,* Appendix 4.

94. Watson, *The Annotated and Illustrated Double Helix,* Appendix 4.

95. Aaron Klug, "Rosalind Franklin and the Discovery of the Structure of DNA", *Nature* 219, no. 5156 (1968): 808; Anne Sayre, *Rosalind Franklin and DNA* (New York: Norton, 1975); and Brenda Maddox, *Rosalind Franklin: The Dark Lady of DNA* (New York: Harper Perennial, 2003).

96. Watson, *The Annotated and Illustrated Double Helix*, 10.

97. Watson, *The Annotated and Illustrated Double Helix*, 12.

98. Doudna and Sternberg, *A Crack in Creation*, 11.

99. See discussion in Chapter 1. Lederman refers to this myth-historical "truth" in Lederman with Teresi, *The God Particle*, 412.

100. Doudna and Sternberg, *A Crack in Creation*, 39.

101. Doudna and Sternberg, *A Crack in Creation*, 61.

102. Doudna and Sternberg, *A Crack in Creation*, 61–62.

103. Doudna and Sternberg, *A Crack in Creation*, 91.

104. Doudna and Sternberg, *A Crack in Creation*, 91.

105. Doudna and Sternberg, *A Crack in Creation*, 96–97.

106. Mertz reflecting on Lander's history: "This is about science politics, not science." As discussed earlier, Mertz herself was erased from the development of rDNA, and yet does not see this as a systematic part of science (quoted in Kozubek, *Modern Prometheus,* 29–30).

107. Doudna and Sternberg, *A Crack in Creation*, 92.

108. Doudna and Sternberg, *A Crack in Creation*, 89.

109. Doudna and Charpentier became the first all-female team to win the Nobel Prize in Chemistry. For more on this, see Alice Park, "Nobel Prize in Chemistry Awarded to First All-Female Team for CRISPR Gene Editing," *Time Magazine*, October 8, 2020. https://time.com/5897538/nobel-prize-crispr-gene-editing/.

110. Doudna and Sternberg, *A Crack in Creation*, 117–133.

111. Doudna and Sternberg, *A Crack in Creation*, 139–141.

112. Doudna and Sternberg, *A Crack in Creation*, 66.

113. Doudna was a cofounder of three major CRISPR companies. Besides Caribou, she cofounded Editas Medicine with Zhang, Church, and others from Cambridge. When the patent dispute heated up, she left Editas and cofounded a competitor based in California—Intellia Therapeutics. For more on this, see Cohen, "How the Battle Lines over CRISPR Were Drawn" and Doudna and Sternberg, *A Crack in Creation*, 66.

114. Doudna and Sternberg, *A Crack in Creation*, 97.

115. Doudna and Sternberg, *A Crack in Creation*, 113.

116. Doudna and Sternberg, *A Crack in Creation*, 119.

117. Doudna and Sternberg, *A Crack in Creation*, 117.

118. Doudna and Sternberg, *A Crack in Creation*, 125.

119. Doudna and Sternberg, *A Crack in Creation*, 120.
120. Doudna and Sternberg, *A Crack in Creation*, 200.
121. Doudna and Sternberg, *A Crack in Creation*, 160.
122. Doudna and Sternberg, *A Crack in Creation*, 160.
123. Doudna and Sternberg, *A Crack in Creation*, 233–234.
124. Doudna and Sternberg, *A Crack in Creation*, 119.
125. Doudna and Sternberg, *A Crack in Creation*, 244.
126. Doudna and Sternberg, *A Crack in Creation*, 244.
127. Doudna and Sternberg, *A Crack in Creation*, 193.
128. William D. Stansfield, "Luther Burbank: Honorary Member of the American Breeders' Association," *Journal of Heredity* 97, no. 2, (2006): 95–99. https://academic.oup.com/jhered/article/97/2/95/2187603.
129. For discussion of Feynman's visit of the 1933 World's Fair in Chicago, see Chapter 1.
130. As quoted in Daniel J. Kevles, *In the Name of Eugenics: Genetics and the Uses of Human Heredity* (New York: Knopf Doubleday, 1985), Kindle edition locations, 314–315.
131. Montreal *Gazette*. "Sterilization of Unfit Advocated: Feeble-Minded Increasing at Disproportionate Rate in Canada." *The Montreal Gazette*, November 14, 1933. https://news.google.com/newspapers?id=oBYvAAAAIBAJ&pg=6315,1778929.
132. For an excellent analysis of the rehabilitation of eugenics, see Gordin, *The Pseudoscience Wars*, 106–134.
133. Antonio Regalado, "Exclusive: Chinese scientists Are Creating CRISPR Babies," *MIT Technology Review*, November 25, 2018. https://www.technologyreview.com/2018/11/25/138962/exclusive-chinese-scientists-are-creating-crispr-babies/; Todd Ackerman, "Chinese Scientist, Assisted by Rice Professor, Claims First Gene-Edited Babies," *AP News*, November 26, 2018. https://apnews.com/94fecf56cac841639d3bc9bad8db1698.
134. Ackerman, "Chinese Scientist, Assisted by Rice Professor, Claims First Gene-Edited Babies."
135. Rob Stein, "New U.S. Experiments Aim to Create Gene-Edited Human Embryos," National Public Radio, February 1, 2019. https://www.npr.org/sections/health-shots/2019/02/01/689623550/new-u-s-experiments-aim-to-create-gene-edited-human-embryos.

Chapter 4: Echoes of Gravitational Waves

1. Walt Whitman, *Walt Whitman: The Complete Poems*, ed. Francis Murphy (London: Penguin Books, 2004), 123.

2. Quoted in Richard S. Westfall, *Never at Rest: A Biography of Isaac Newton* (Cambridge: Cambridge University Press, 1983), 274. Newton's original text was "If I have seen further, it is by standing on ye sholders of Giants."

3. Robert K. Merton, *On the Shoulders of Giants: A Shandean Postscript* (Chicago: University of Chicago Press, 1993).

4. Ahn et al., "Even Einstein Struggled," 314–328.

5. France Córdova at NSF and LIGO Press Conference: National Science Foundation (NSF), "LIGO Detects Gravitational Waves— Announcement at Press Conference," *NSF Press Conference*, video 1:11:29, February 11, 2016. https://www.ligo.caltech.edu/video/ligo20160211v11, 1:00–1:11.

6. David Reitze at NSF and LIGO Press Conference: NSF, "LIGO Detects Gravitational Waves," 4:02–4:22.

7. Spacetime refers to a four-dimensional manifold sometimes referred to as the "fabric" of which our universe is made. The concept grew out of Einstein's special theory of relativity. To pinpoint an object you need three coordinates of space (x, y, and z) plus a measurement of time. The theory of general relativity describes what happens when energy and mass cause spacetime to warp or curve. Forces like gravity can be understood as our experience of the curvature of spacetime. Gravitational waves arise from sharp disturbances in spacetime from extremely dense (massive) objects. For a clear, nontechnical introduction to the concept of spacetime, see David J. Griffiths, *Revolutions in Twentieth-Century Physics* (Cambridge: Cambridge University Press, 2013), Sections 2.5 and 5.4.2.

8. LIGO Scientific Collaboration (LSC), "Calibration of the Advanced LIGO Detectors for the Discovery of the Binary Black-Hole Merger GW150914," LIGO.org. https://www.ligo.org/science/Publication-GW150914Calibration/index.php.

9. France Córdova at NSF and LIGO Press Conference: NSF, "LIGO Detects Gravitational Waves," 1:13–1:37.

10. A laser interferometer is a high-precision measuring device. It works by merging two sources of laser light and studying the resulting interference pattern. For more information, see LIGO Caltech, "What Is an Interferometer?" LIGO Caltech—Learn More. https://www.ligo.caltech.edu/page/what-is-interferometer.

11. David Reitze at NSF and LIGO Press Conference: NSF, "LIGO Detects Gravitational Waves," 10:15–10:30.

12. David Reitze showing simulations at NSF and LIGO Press Conference: NSF, "LIGO Detects Gravitational Waves," 6:25–9:56.

13. Adrian Cho, "Here's the First Person to Spot Those Gravitational Waves," *ScienceMag*, February 11, 2016. https://www.sciencemag.org/news/2016/02/here-s-first-person-spot-those-gravitational-waves.

14. Reitze at NSF and LIGO Press Conference: NSF, "LIGO Detects Gravitational Waves," 4:02–4:22.

15. Rainer Weiss at NSF and LIGO Press Conference: NSF, "LIGO Detects Gravitational Waves," 21:42.

16. In this chapter, Einstein's general theory of relativity will be denoted GTR; general relativity as a field of study will be denoted GR.

17. Weiss at NSF and LIGO Press Conference: NSF, "LIGO Detects Gravitational Waves," 22:40–23:06.

18. Weiss at NSF and LIGO Press Conference: NSF, "LIGO Detects Gravitational Waves," 24:15–25:06.

19. Kip Thorne at NSF and LIGO Press Conference: NSF, "LIGO Detects Gravitational Waves," 31:55–33:03. This same linear progress narrative is laid out in a bit more detail in the peer-reviewed article published that same day announcing the discovery. B.P. Abbot et al., "Observation of Gravitational Waves from a Binary Black Hole Merger." *Physical Review Letters* 116, 061102 (2016): 1.

20. France Córdova at NSF and LIGO Press Conference: NSF, "LIGO Detects Gravitational Waves," 40:02–42:07.

21. Córdova at NSF and LIGO Press Conference: NSF, "LIGO Detects Gravitational Waves," 42:24.

22. As quoted in LIGO Caltech, "Gravitational Waves Detected 100 Years after Einstein's Prediction," press release, February 11, 2016. https://www.ligo.caltech.edu/news/ligo20160211.

23. The irony is not lost on the author that by studying this particular case, I, too, tap into Einstein's celebrity legacy.

24. Gerald Holton, "Who Was Einstein? Why Is He Still So Alive?," in *Einstein for the 21st Century: His Legacy in Science, Art, and Modern Culture*, eds. Peter L. Galison, Gerald Holton, and Silvan S. Schweber (Princeton: Princeton University Press, 2008).

25. For example, the library at Vassar College has a limited but fascinating collection of Einstein materials. I teach a seminar on Einstein in which we routinely explore this archive. The collection is digitized, searchable, and publicly available; see Vassar College Libraries, "The Albert Einstein Digital Collection at Vassar College Libraries." https://einstein.digitallibrary.vassar.edu.

26. There are numerous well-researched biographical sketches of Einstein. For general biographical information, I relied mostly on Jürgen Neffe, *Einstein: A Biography*, transl. Shelley Frisch (New York:

Farrar, Straus and Giroux, 2007); Michel Janssen and Christoph Lehner, eds., *The Cambridge Companion to Einstein*, Vol. 1 (Cambridge: Cambridge University Press, 2014); and Walter Isaacson, *Einstein: His Life and Universe* (New York: Simon & Schuster, 2007). In addition, numerous sources speak specifically to aspects of Einstein's life and work that are pertinent to this discussion.

27. Matthew Stanley, *Einstein's War: How Relativity Triumphed amid the Vicious Nationalism of World War I* (New York: Dutton, 2019), 8.
28. Stanley, *Einstein's War*, 378.
29. Isaacson, *Einstein*, 16.
30. For more on this, see Neffe, *Einstein*, 304; see also Isaacson, *Einstein*, 61.
31. For more on the effects of Nazism in Germany and the forced migration of scientists, see Isaacson, *Einstein*, 407.
32. Richard S. Levy, "Political Antisemitism in Germany and Austria 1848–1914," in *Antisemitism: A History*, eds. Albert S. Lindemann and Richard S. Levy (Oxford: Oxford University Press, 2010): 127–131.
33. Isaacson, *Einstein*, 152.
34. See Neffe, *Einstein*, 307 and Isaacson, *Einstein*, 270 and 288–289.
35. Daniel Kennefick, *Traveling at the Speed of Thought: Einstein and the Quest for Gravitational Waves* (Princeton: Princeton University Press, 2007), 23–39.
36. Peter Galison, *Einstein's Clocks, Poincaré's Maps: Empires of Time* (New York: W.W. Norton and Co., 2004), 256. Also see Olivier Darrigol, "The Genesis of the Theory of Relativity," in *Einstein, 1905–2005: Poincaré Seminar 2005*, eds. Thibault Damour, Olivier Darrigol, and Vincent Rivasseau (Berlin: Springer Science & Business Media, 2006), 1–22.
37. Kennefick, *Traveling at the Speed of Thought*, 10.
38. Kennefick, *Traveling at the Speed of Thought*, 39.
39. Kennefick, *Traveling at the Speed of Thought*, 91.
40. Kennefick, *Traveling at the Speed of Thought*, 38–39.
41. Kennefick, *Traveling at the Speed of Thought*, 49–54.
42. Kennefick, *Traveling at the Speed of Thought*, 79.
43. Kennefick, *Traveling at the Speed of Thought*, 81–96.
44. Kennefick, *Traveling at the Speed of Thought*, 81–96.
45. Kennefick, *Traveling at the Speed of Thought*, 104.
46. Kennefick, *Traveling at the Speed of Thought*, 105–107.
47. Alexander Blum, Roberto Lalli, and Jürgen Renn, "The Reinvention of General Relativity," *Isis* 106, no. 3 (September 2015): 613.
48. Blum, Lalli, and Renn, "The Reinvention of General Relativity," 612–613 and 618.
49. Blum, Lalli, and Renn, "The Reinvention of General Relativity," 606.

50. Blum, Lalli, and Renn, "The Reinvention of General Relativity," 612–618.
51. Kennefick, *Traveling at the Speed of Thought*, 106.
52. Nathan Rosen, "On Cylindrical Gravitational Waves," in *Proceedings of the Fiftieth Anniversary Conference, Bern, 11–16 July 1955,* eds. André Mercier and Michael Kervaire (Basel: Birkhäuser, 1956), 171–175. Also see D.C. Robinson, "Gravitation and General Relativity at King's College London," *The European Physical Journal H* 44, no. 3 (2019): 181–270, and Claus Kiefer, "Space and Time 62 Years after the Berne Conference," Preprint. Accessed February 18, 2020: https://arxiv.org/pdf/1812.10679.pdf, 5.
53. Kennefick, *Traveling at the Speed of Thought*, 211–224.
54. Dean Rickles, "The Chapel Hill Conference in Context," in *The Role of Gravitation in Physics: Report from the 1957 Chapel Hill Conference*, eds. D. Rickles and C.M. DeWitt. Communicated by Jürgen Renn, Alexander Blum, and Peter Damerow, Edition Open Access, 2011: https://edition-open-sources.org/sources/5/, 59.
55. Rickles, "The Chapel Hill Conference in Context," 7–21.
56. Rickles, "The Chapel Hill Conference in Context," 29 and original front material.
57. Feynman actually showed up late to the conference as he recounts in *Surely You're Joking, Mr. Feynman!* When he arrived at the airport in North Carolina, he had no idea where the conference was being held, and had to do some sleuthing with taxi drivers. See Richard P. Feynman and Ralph Leighton, *"Surely You're Joking, Mr. Feynman!" Adventures of a Curious Character* (New York: Random House, 1992), 258–259.
58. D. Rickles and C.M. DeWitt, eds., *The Role of Gravitation in Physics: Report from the 1957 Chapel Hill Conference*. Communicated by Jürgen Renn, Alexander Blum, and Peter Damerow, Edition Open Access, 2011. https://edition-open-sources.org/sources/5/, 243–260.
59. Felix A.E. Pirani, "Republication of: On the Physical Significance of the Riemann Tensor," *General Relativity Gravitation* 41 (2009): 1215–1232. Also see Rickles and DeWitt eds., *The Role of Gravitation in Physics*, 279–281.
60. Hermann Bondi, "Plane Gravitational Waves in General Relativity," *Nature* 179 (1957): 1072–1073.
61. Marcia Bartusiak, *Einstein's Unfinished Symphony: The Story of a Gamble, Two Black Holes, and a New Age of Astronomy* (New Haven: Yale University Press, 2017), 96.
62. Virginia Trimble, "Wired by Weber: The Story of the First Searcher and Searches for Gravitational Waves," *The European Physical Journal*

H 42, no. 2 (2017): 262–263. Also see interview of Joseph Weber by Joan Bromberg on April 8, 1983, Niels Bohr Library & Archives, American Institute of Physics (AIP), Oral Histories, College Park, MD. https://www.aip.org/history-programs/niels-bohr-library/oral-histories/4941.

63. Weber, interview by Joan Bromberg, AIP Oral Histories.
64. Weber, interview by Joan Bromberg, AIP Oral Histories.
65. Weber, interview by Joan Bromberg, AIP Oral Histories. Also see Trimble, "Wired by Weber," 262–263.
66. Weber, interview by Joan Bromberg, AIP Oral Histories.
67. Weber, interview by Joan Bromberg, AIP Oral Histories.
68. Mario Bertolotti, *The History of the Laser* (London: Institute of Physics—IOP, 2005), 179–181. Also see Trimble, "Wired by Weber," 262–263. Although both MASER and LASER are acronyms, I adopt the more conventional use of lowercase letters: maser and laser.
69. Bertolotti, *The History of the Laser*, 179–188.
70. Collins, *Gravity's Shadow*, 25. Also see Trimble, "Wired by Weber," 264.
71. Kip S. Thorne, *Black Holes and Time Warps: Einstein's Outrageous Legacy* (New York: W.W. Norton & Co., 1995), 366–367.
72. According to Trimble, during that sabbatical Weber was affiliated with both the IAS and Princeton University. See Trimble, "Wired by Weber," 262–263 and 270.
73. Trimble, "Wired by Weber," 262–263 and 270.
74. Trimble, "Wired by Weber," 262–263 and 270.
75. Thorne, *Black Holes and Time Warps*, 366–367.
76. Joseph Weber and John Archibald Wheeler, "Reality of the Cylindrical Gravitational Waves of Einstein and Rosen," *Reviews of Modern Physics* 29 (1957): 509–515. Also see Kennefick, *Traveling at the Speed of Thought*, 135–137.
77. Kennefick, *Traveling at the Speed of Thought*, 139–141.
78. Thorne, *Black Holes and Time Warps*, 369–370. Also see Trimble, "Wired by Weber," 262–263.
79. Thorne, *Black Holes and Time Warps*, 367.
80. Collins, *Gravity's Shadow*, 31–32. Also see Weber's original *Physical Review* paper: "Detection and Generation of Gravitational Waves," *Physical Review Letters* 117 (1960), 306.
81. Bartusiak, *Einstein's Unfinished Symphony*, 100.
82. Allan Franklin, "How to Avoid the Experimenters' Regress," *Studies in History and Philosophy of Science* 25, no. 3 (1994): 466.
83. Bartusiak, *Einstein's Unfinished Symphony*, 100.
84. Kennefick, *Traveling at the Speed of Thought*, 123.

85. See references in Trimble, "Wired by Weber," 288–289.

86. Kennefick. *Traveling at the Speed of Thought,* 123.

87. Weber's 1969 paper contains a table showing coincidence counts. Nine were two-detector coincidences, five were coincident with three-detectors, and three were coincident on all four detectors taking data at the time. See Collins, *Gravity's Shadow,* 85–86.

88. Bartusiak, *Einstein's Unfinished Symphony,* 100.

89. Trimble, "Wired by Weber," 262–263 and 274. Also see Allan Franklin, "Gravity Waves and Neutrinos: The Latter work of Joseph Weber," *Perspectives in Science* 18, no. 2 (2010): 121.

90. Bartusiak, *Einstein's Unfinished Symphony,* 100–101.

91. Harry Collins, *Gravity's Kiss: The Detection of Gravitational Waves* (Cambridge, MA: MIT Press, 2017), 50.

92. Collins and Pinch, *The Golem,* 91.

93. Collins, *Gravity's Shadow,* 13–14 and 84–85. Collins uses a hyphenated form of this concept "social space-time," For consistency, I use the unhyphenated form.

94. See Franklin, "How to Avoid the Experimenters' Regress," 463–491, and Franklin, "Gravity Waves and Neutrinos," 119–151.

95. Franklin and Collins, "Two Kinds of Case Study and a New Agreement," 88–115.

96. Franklin, "Gravity Waves and Neutrinos," 121.

97. Franklin, "How to Avoid the Experimenters' Regress," 467–468.

98. William H. Press and Kip S. Thorne, "Gravitational-Wave Astronomy," *Annual Review of Astronomy and Astrophysics* 10, no. 1 (1972): 335.

99. Collins, *Gravity's Shadow,* 126. In his analysis, Franklin explicitly denies the reality of an experimenter's regress, and thinks Collins's reading of this controversy is distortive. See Franklin, "How to Avoid the Experimenters' Regress," 463–491. Though I agree in large part with Franklin's argument that the epistemological case for rejecting Weber's claims of detection was reasonable, I also agree with Collins's conclusion that what pushed the community into consensus on this point included social dynamics that venture beyond Mertonian norms of science.

100. Collins, *Gravity's Shadow,* 126–128.

101. Collins, *Gravity's Shadow,* 126–153. Here again Franklin diverges from Collins, as he stresses the intergroup collaborations between various critics. However, I find Collins's evidence on this point very convincing.

102. Collins, *Gravity's Shadow,* 153.

103. This is a critical point of departure between Collins and Franklin. Collins argues that to break the experimenter's regress, social forces were used: no one attempted to replicate Weber's experiment and therefore his results were not explicitly disqualified. Franklin, on the other hand, argues that there was enough evidence from various sources that discredited Weber's results to make the consensus evidence-based and reasonable.

104. Collins, *Gravity's Shadow*, 154–188.

105. Collins, *Gravity's Shadow*, 169–170.

106. Collins, *Gravity's Shadow*, 173–188.

107. Collins, *Gravity's Shadow*, 164.

108. Collins, *Gravity's Shadow*, 155.

109. Collins and Pinch, *The Golem*, 99.

110. Collins, *Gravity's Shadow*, 155.

111. This wasn't completely true, however: both Collins and Franklin show that when Weber ventured out of gravitational waves and into neutrino detection, his work faced less hostility. See Franklin, "Gravity Waves and Neutrinos," 119–151.

112. Collins, *Gravity's Shadow*, 196.

113. Adrian Cho, "Remembering Joseph Weber, the Controversial Pioneer of Gravitational Waves," *ScienceMag*, February 12, 2016. https://www.sciencemag.org/news/2016/02/remembering-joseph-weber-controversial-pioneer-gravitational-waves.

114. Collins, *Gravity's Shadow*, 358–391.

115. Collins, *Gravity's Shadow*, 360–370.

116. According to Collins's interviews and analysis, Weber's loss of funding was not inevitable. However, the NSF made a calculation and chose to cut his funding to avoid the possibility of future problems Weber might cause for LIGO. See Collins, *Gravity's Shadow*, 358–391.

117. Collins, *Gravity's Shadow*, 358–391.

118. Cho, "Remembering Joseph Weber."

119. According to Trimble, although Weber is a pioneer of gravitational wave detection, the American Institute of Physics has no audio recordings or interviews with Weber about his work on gravitational waves. See Toni Feder, "Q&A: Virginia Trimble on 50-plus Years in Astronomy," *Physics Today*, March 13, 2018. https://physicstoday.scitation.org/do/10.1063/pt.6.4.20180313a/full/.

120. Feder, "Q&A: Virginia Trimble on 50-plus Years in Astronomy." Also see Trimble, "Wired by Weber," 266.

121. Trimble notes that she and Weber had friendly relationships with many important scientists, including Feynman. She also says that

she has been friends with Córdova for decades. See Trimble, "Wired by Weber," 281 and 283.

122. Kip S. Thorne, "The Generation of Gravitational Waves: A Review of Computational Techniques," in *Topics in Theoretical and Experimental Gravitation Physics*, eds. V. De Sabbata and J. Weber (New York: Plenum Press, 1977), 1–61.

123. Thorne, *Black Holes and Time Warps,* 366–370.

124. Collins, *Gravity's Shadow*, 811.

125. Trimble describes seeing Robert Forward's first interferometer in 1971. See Trimble, "Wired by Weber," 271–272.

126. Franklin, "Gravity Waves and Neutrinos," 147.

Chapter 5: Demarcating Seismic Uncertainties

1. George Orwell, *Nineteen Eighty-Four* (New York: Harcourt, Brace and Company, 1949), 74–75.

2. For more on the theme of candle in the darkness, see Carl Sagan, *Demon-Haunted World: Science as a Candle in the Dark* (New York: Ballantine Books, 1996).

3. Lederman with Teresi, *The God Particle,* 412 and 197.

4. A great example of this is the use of the von Neumann impossibility proof against hidden variables discussed in Chapter 2.

5. For a broader discussion on this topic, see Nate Silver, *The Signal and The Noise: Why So Many Predictions Fail—But Some Don't* (New York: Penguin Books, 2015); Monya Baker, "1,500 Scientists Lift the Lid on Reproducibility," *Nature News* 533, no. 7604 (2016): 452–454; M. Munafò, B. Nosek, D. Bishop, et al. "A Manifesto for Reproducible Science," *Nature Human Behaviour* 1, no. 0021 (2017): 1–9. https://www.nature.com/articles/s41562-016-0021; Brian Keating, "How My Nobel Dream Bit the Dust," *Nautilus*, Connections no. 059, April 19, 2018. http://nautil.us/issue/59/connections/how-my-nobel-dream-bit-the-dust; and Jonathan P. Tennant, Jonathan M. Dugan, Daniel Graziotin, et al. "A Multi-disciplinary Perspective on Emergent and Future Innovations in Peer Review," *F1000Research* 6 (2017). https://f1000research.com/articles/6-1151.

6. Munafò, Nosek, Bishop, et al. "A Manifesto for Reproducible Science."

7. 52% of respondents thought it was a "significant crisis" while 38% thought it was a "slight crisis." See Baker, "1,500 Scientists Lift the Lid on Reproducibility," 452. In some literature the crisis is referred to as the "replication" crisis, but I will use the term "reproducibility."

8. Baker, "1,500 Scientists Lift the Lid on Reproducibility," 452.
9. A good example of this is the BICEP2 controversy discussed in the Conclusion of this book. See Keating, "How My Nobel Dream Bit the Dust." This short memoir of the BICEP2 drama was adapted from Keating, *Losing the Nobel Prize: A Story of Cosmology, Ambition, and the Perils of Science's Highest Honor* (New York: W.W. Norton & Co., 2018).
10. Valentin Amrhein, David Trafimow, and Sander Greenland, "Inferential Statistics as Descriptive Statistics: There Is No Replication Crisis if We Don't Expect Replication," *The American Statistician* 73, Supplement 1 (2019): 262.
11. See the BICEP2 controversy discussed in the Conclusion. Keating, "How My Nobel Dream Bit the Dust."
12. Silver, *The Signal and the Noise,* 3–4.
13. Silver, *The Signal and the Noise,* 163–171.
14. See discussion in Chapter 1 exploring Einstein's quote from his *Essays.*
15. Shawn Otto, "A Plan to Defend Against the War on Science," *Scientific American* 9, October 9, 2016. https://www.scientificamerican.com/article/a-plan-to-defend-against-the-war-on-science/. For a more complete analysis, see Otto, *The War on Science.*
16. Otto, *The War on Science.* See also Oreskes and Conway, *Merchants of Doubt*; Oreskes, *Why Trust Science?*; and Michaels, *The Triumph of Doubt.*
17. Alice Benessia and Bruna De Marchi, "When the Earth Shakes . . . and Science with It. The Management and Communication of Uncertainty in the L'Aquila Earthquake," *Futures* 91 (2017): 41.
18. This reading corresponds to the moment magnitude scale (M_w) as opposed to the older, generally more familiar, Richter magnitude scale (M_l).
19. T.H. Jordan et al., "Operational Earthquake Forecasting: State of Knowledge and Guidelines for Implementation, Final Report of the International Commission on Earthquake Forecasting for Civil Protection," *Annals of Geophysics* 54, no. 4 (2011): 320–321.
20. Benessia and De Marchi, "When the Earth Shakes," 35.
21. D.E. Alexander, "Communicating Earthquake Risk to the Public: The Trial of the 'L'Aquila Seven'," *Journal of the International Society for the Prevention and Mitigation of Natural Hazards* 72, no. 2 (2014): 1160.
22. As quoted in Benessia and De Marchi, "When the Earth Shakes," 41. Quote from Judge Billi, "Tribunale di L'Aquila, Sezione Penale. Motivazione Sentenza n. 380 del 22/10/2012, Depositata il 19/01/2013." http://www.magistraturademocratica.it/mdem/qg/doc/Tribunale_di_LAquila_sentenza_condanna_Grandi_Rischi_terremoto.pdf. See also Edwin Cartlidge, "Italy's Supreme Court Clears L'Aquila

Earthquake Scientists for Good," *ScienceMag News,* November 20, 2015. https://www.sciencemag.org/news/2015/11/italy-s-supreme-court-clears-l-aquila-earthquake-scientists-good, and David Ropeik, "The L Aquila Verdict: A Judgment Not against Science, but against a Failure of Science Communication," *Scientific American175,* October 22, 2012. https://blogs.scientificamerican.com/guest-blog/the-laquila-verdict-a-judgment-not-against-science-but-against-a-failure-of-science-communication/.

23. Povoledo, E. and H. Fountain, "Italy Orders Jail Terms for 7 Who Didn't Warn of Deadly Earthquake," *The New York Times,* October 22, 2012. https://www.nytimes.com/2012/10/23/world/europe/italy-convicts-7-for-failure-to-warn-of-quake.html.

24. Letter from Alan Leshner (CEO of AAAS) to Giorgio Napolitano (President of the Republic of Italy), June 29, 2010. Alan I. Leshner, "Letter to the President of the Italian Republic," American Association for the Advancement of Science (AAAS), June 30, 2010. http://www.aaas.org/sites/default/files/migrate/uploads/0630italy_letter.pdf.

25. Leshner, "Letter to the President."

26. Leshner, "Letter to the President."

27. T. Kington, "L'Aquila's Earthquake-Scarred Streets See Battle between Science and Politics," *The Guardian,* October 27, 2012. https://www.theguardian.com/world/2012/oct/27/laquila-earthquake-battle-science-politics.

28. P. Pietrucci, "Voices from the Seismic Crater in the Trial of the Major Risk Committee: A Local Counternarrative of 'the L'Aquila Seven,'" *Interface* 8, no. 2 (2016): 270; Stephen S. Hall, "Scientists on Trial: At Fault?," *Nature News,* September 14, 2011. https://www.nature.com/news/2011/110914/full/477264a.html.

29. The Dipartimento della Protezione Civile (DPC) is an Italian national government agency tasked to manage and protect public safety. From here on, it will be referred to as the DPC.

30. Hall, "Scientists on Trial: At Fault?," 266.

31. Hall, "Scientists on Trial: At Fault?," 269.

32. Hall, "Scientists on Trial: At Fault?," 265.

33. Danielle DeVasto et al., "Stasis and Matters of Concern: The Conviction of the L'Aquila Seven," *Journal of Business and Technical Communication* 30, no. 2 (2016): 141.

34. Pietrucci, "Voices from the Seismic Crater," 272.

35. Pietrucci, "Voices from the Seismic Crater," 273. The National Institute of Geophysics and Volcanology (INGV) has a national

earthquake center that lists real-time seismic data on its website. See Istituto Nazionale di Geofisica e Vulcanologia (INGV), "Lista Terremoti," INGV.it. http://terremoti.ingv.it. TVUNO is one of the main television channels in Italy, and a source of breaking news.

36. J. Dollar, "The Man Who Predicted an Earthquake," *The Guardian*, April 5, 2010. https://www.theguardian.com/world/2010/apr/05/laquila-earthquake-prediction-giampaolo-giuliani.

37. Hall, "Scientists on Trial: At Fault?," 265.

38. Edwin Cartlidge, "Updated: Appeals Court Overturns Manslaughter Convictions of Six Earthquake Scientists," *Science News*, November 10, 2014. https://www.sciencemag.org/news/2014/11/updated-appeals-court-overturns-manslaughter-convictions-six-earthquake-scientists.

39. Leshner, "Letter to the President."

40. See Susan E. Hough, *Predicting the Unpredictable: The Tumultuous Science of Earthquake Prediction* (Princeton: Princeton University Press, 2016) and Helene Joffe et al., "Stigma in Science: The Case of Earthquake Prediction," *Disasters* 42, no. 1 (2018): 81–83.

41. Jordan et al., "Operational Earthquake Forecasting," 321.

42. See for example United States Geological Survey (USGS), "Can You Predict Earthquakes?," USGS.gov. https://www.usgs.gov/faqs/can-you-predict-earthquakes?qt-_news_science_products=0&qt-news_science_products=0#qt-news_science_products.

43. Mallet wrote an introductory essay for a book by Luigi Palmieri, *The Eruption of Vesuvius in 1872* (London: Asher & Company, 1873), 44–45.

44. Naomi Oreskes, "From Continental Drift to Plate Tectonics," in *Plate Tectonics: An Insider's History of the Modern Theory of the Earth*, ed. Naomi Oreskes (London: CRC Press, 2018).

45. Bruce A. Bolt, "Locating Earthquakes and Plate Boundaries," in *Plate Tectonics: An Insider's History of the Modern Theory of the Earth*, ed. Naomi Oreskes (London: CRC Press, 2018). In the same volume, see Jack Oliver, "Earthquake Seismology in the Plate Tectonics Revolution."

46. Elizabeth Kolbert, "The Shaky Science behind Predicting Earthquakes," *Smithsonian Magazine*, June 2015. https://www.smithsonianmag.com/science-nature/shaky-science-behind-predicting-earthquakes-180955296/.

47. Nate Silver, *The Signal and the Noise*, 146–148.

48. Susan E. Hough, "Earthquakes: Predicting the Unpredictable?," *Geotimes* 50, no. 3 (March 2005). http://www.geotimes.org/mar05/feature_eqprediction.html.

49. Kolbert, "The Shaky Science behind Predicting Earthquakes."

50. Hough, "Earthquakes: Predicting the Unpredictable?"

51. USGS, "Can You Predict Earthquakes?"

52. USGS, "Can You Predict Earthquakes?"

53. USGS, "Can You Predict Earthquakes?"

54. See, for example, Robert J. Geller, "Earthquake Prediction: A Critical Review," *Geophysical Journal International* 131, no. 3 (1997): 425–450; G. Igarashi et al., "Ground-Water Radon Anomaly before the Kobe Earthquake in Japan," *Science* 269, no. 5220 (1995): 60–61; Vivek Walia et al., "Earthquake Prediction Studies Using Radon as a Precursor in NW Himalayas, India: A Case Study," *TAO: Terrestrial, Atmospheric and Oceanic Sciences* 16, no. 4 (2005): 775–804.

55. Rachel A. Grant and Tim Halliday, "Predicting the Unpredictable; Evidence of Pre-Seismic Anticipatory Behaviour in the Common Toad." *Journal of Zoology* 281, no. 4 (2010): 263–271.

56. Research at Los Alamos National Laboratory uses machine learning to tease out a signal from seismic noise. Is the pendulum now swinging back to optimism? For more on this, see Paul Johnson, "The Hidden Seismic Symphony in Earthquake Signals," *Scientific American175,* April 25, 2019. https://blogs.scientificamerican.com/observations/the-hidden-seismic-symphony-in-earthquake-signals/.

57. Jordan et al., "Operational Earthquake Forecasting," 321.

58. A seismic "swarm" is defined by the USGS as "a sequence of mostly small earthquakes with no identifiable mainshock. Swarms are usually short-lived, but they can continue for days, weeks, or sometimes even months." For more on this, see USGS, "What Is the Difference between Aftershocks and Swarms?" https://www.usgs.gov/faqs/what-difference-between-aftershocks-and-swarms?qt-news_science_products=0#qt-news_science_products.

59. David E. Alexander, "The L'Aquila Earthquake of 6 April 2009 and Italian Government Policy on Disaster Response," *Journal of Natural Resources Policy Research* 2, no. 4 (October 2010): 327.

60. Jordan et al., "Operational Earthquake Forecasting," 325.

61. Alexander, "The L'Aquila Earthquake," 326.

62. Kolbert, "The Shaky Science behind Predicting Earthquakes."

63. Hall, "Scientists on Trial: At Fault?," 268.

64. A master's thesis by Dario del Fante, who grew up in L'Aquila, is a firsthand account and a media analysis of the period leading up to the earthquake. Del Fante's thesis includes references to, and quotes from, articles in *Il Centro* (The Center) newspaper as well as first-person testimonials. All subsequent references to newspaper accounts are from his analysis. Dario del Fante, "In Margine al

Terremoto de L'Aquila: Il Caso Giuliani" (Master's thesis, University of Siena, 2013). https://www.academia.edu/13446373/in_margine_al_terremoto_de_l_aquila_il_caso_giuliani.

65. Del Fante, "In Margine al Terremoto de L'Aquila."

66. The National Institute of Geophysics and Volcanology (Istituto Nazionale di Geofisica e Vulcanologia, INGV), in Rome, is funded by the Italian government and is responsible for coordinating research, education, and outreach on seismic and volcanic activity in the country.

67. As quoted in del Fante, "In Margine al Terremoto de L'Aquila," 13. From the online archive of *Il Centro* newspaper in L'Aquila, January 25, 2009.

68. As quoted in del Fante, "In Margine al Terremoto de L'Aquila."

69. Pietrucci, "Voices from the Seismic Crater," 272.

70. Benessia and De Marchi, "When the Earth Shakes," 38.

71. An interesting question arises when considering the saying coined by the seismologist Nicolas Ambraseys, "Earthquakes don't kill people; buildings kill people." See Hough, *Predicting the Unpredictable*, vii-viii. Why did the *Aquilani* feel betrayed by science, not by the local building inspectors? Most of the dead were crushed under buildings that should have been built to withstand seismic shocks like the one of 2009. One answer might be that the *Aquilani* were particularly angry with scientists due to their proactive, distortive rhetoric that could be tied to specific people at specific times. But the lack of structural integrity in the buildings was a passive, ongoing, and communal negligence.

72. The LVD is one of the largest neutrino observatories in the world. It is part of Italy's Istituto Nazionale Di Fisica Nucleare (INFN). See Istituto Nazionale Di Fisica Nucleare (INFN), "Laboratori Nazionali del Gran Sasso (LNGS)—LVD," INFN.it. https://www.lngs.infn.it/en/lvd.

73. Istituto Nazionale di Fisica Nucleare (INFN), "Laboratori Nazionali del Gran Sasso (LNGS)—ERMES," INFN.it. https://www.lngs.infn.it/en/ermes.

74. See, for example, Geller, "Earthquake Prediction," 425–450; Igarashi et al., "Ground-Water Radon Anomaly," 60–61; and Walia et al., "Earthquake Prediction Studies Using Radon," 775–804.

75. Walia et al., "Earthquake Prediction Studies Using Radon," 776.

76. Although mainstream seismologists may disregard it, there is no scientific prohibition against this line of research. In 2010, a team of physicists led by Nobel laureate Georges Charpak was building a radon detection system that could be used for earthquake prediction.

James Dacey, "A Radon Detector for Earthquake Prediction," *Physicsworld,* March 18, 2010. https://physicsworld.com/a/a-radon-detector-for-earthquake-prediction/#:~:text=Now%2C%20however%2C%20a%20group%20of,zones%20prior%20to%20earth%20slipping.

77. On Giuliani scientific technician turned earthquake hunter, see Dollar, "The Man Who Predicted an Earthquake."

78. It should be noted that the Molise earthquake was more than 200 km from Giuliani's detectors in L'Aquila. There is no corroborating evidence that he actually predicted it. Alexander, "Communicating Earthquake Risk to the Public," 1161.

79. Just over US $86,000 in 2009. From Italian magazine *Club 3*: Di Neda Accili, "Qui C'è Aria Di Terremoto," *Club 3*, (October, 2008): 50–51. http://www.stpauls.it/club3/0810c3/1008Art_Inventore.pdf.

80. Hall, "Scientists on Trial: At Fault?," 267.

81. Jordan et al., "Operational Earthquake Forecasting," 323.

82. Dollar, "The Man Who Predicted an Earthquake."

83. Alexander, "The L'Aquila Earthquake," 330.

84. Richard A. Kerr, "After the Quake, in Search of the Science—or Even a Good Prediction," *Science* 324 (April 2009): 322, and Alexander, "The L'Aquila Earthquake," 330.

85. Alexander, "The L'Aquila Earthquake," 330. The DPC is short for Dipartimento della Protezione Civile or Department of Civil Protection, an Italian national government agency charged with managing and protecting public safety.

86. Alexander, "Communicating Earthquake Risk to the Public," 1162 and "The L'Aquila Earthquake," 325–342.

87. Ian Sample, "Scientist Was Told to Remove Internet Prediction of Italy Earthquake," *The Guardian* (April 6, 2009). https://www.the-guardian.com/world/2009/apr/06/italy-earthquake-predicted, and Dollar, "The Man Who Predicted an Earthquake."

88. The Commissione Nazionale per la Previsione e Prevenzione dei Grandi Rischi (National Commission for the Prediction and Prevention of Major Risks) is also widely known as the Commissione Grandi Rischi (CGR).

89. DeVasto et al., "Stasis and Matters of Concern," 141–142.

90. DeVasto et al., "Stasis and Matters of Concern," 142.

91. DeVasto et al., "Stasis and Matters of Concern," 142.

92. DeVasto et al., "Stasis and Matters of Concern," 143.

93. As quoted and translated in Benessia and De Marchi, "When the Earth Shakes," 37 and DeVasto et al., "Stasis and Matters of Concern," 143.

94. DeVasto et al., "Stasis and Matters of Concern," 143.
95. DeVasto et al., "Stasis and Matters of Concern," 161 and Hall, "Scientists on Trial: At Fault?," 265.
96. For a full accounting of his academic background, see Bernardo De Bernardinis, "Curriculum Vitae di Bernardo De Bernardinis Aggiornato Agosto 2013," ISPRA. https://www.isprambiente.gov.it/files/trasparenza/organizzazione/cda/curriculum_de_bernardinis.pdf/view.
97. DeVasto et al., "Stasis and Matters of Concern," 149–158.
98. DeVasto et al., "Stasis and Matters of Concern," 150–151.
99. DeVasto et al., "Stasis and Matters of Concern," 154.
100. Benessia and De Marchi, "When the Earth Shakes," 37.
101. Benessia and De Marchi, "When the Earth Shakes," 37.
102. Benessia and De Marchi, "When the Earth Shakes," 38.
103. Edwin Cartlidge, "Seven-year Legal Saga Ends as Italian Official is Cleared of Manslaughter in Earthquake Trial," *ScienceMag News*, October 3, 2016. https://www.sciencemag.org/news/2016/10/seven-year-legal-saga-ends-italian-official-cleared-manslaughter-earthquake-trial.
104. Benessia and De Marchi, "When the Earth Shakes," 37.
105. Cartlidge, "Italy's Supreme Court Clears L'Aquila Earthquake Scientists for Good."
106. Letter from Leshner (CEO of AAAS) to Napolitano (President of the Republic of Italy) June 29, 2010; Leshner, "Letter to the President."
107. Leshner, "Letter to the President."
108. Pietrucci, "Voices from the Seismic Crater," 263.
109. Pietrucci, "Voices from the Seismic Crater," 263–264.
110. Cartlidge, "Updated: Appeals Court Overturns Manslaughter Convictions."
111. Cartlidge, "Seven-Year Legal Saga Ends."
112. Baker, "1,500 Scientists Lift the Lid on Reproducibility," 452–454.
113. "P-hacking" refers to statistical manipulation of data in which a researcher selects and parses data during analysis to produce a probability value under 0.05, a measure of statistical significance.
114. Oreskes and Conway, *Merchants of Doubt*. For a more recent account that reinforces and extends much of Oreskes and Conway's arguments, see Michaels, *The Triumph of Doubt*.
115. Sagan, *Demon-Haunted World*, 26.

Conclusion: Beyond the Wars on Science

1. Adichie, "The Danger of a Single Story," video: 17:32.
2. Morris, *The Ashtray*, 2.
3. Morris, *The Ashtray*, 27.
4. Reviewers of Morris's book have commented on his distorted portrayal of Kuhn's ideas. In particular, see Philip Kitcher, "The Ashtray Has Landed: The Case of Morris v. Kuhn," review of *The Ashtray* by Errol Morris, *Los Angeles Review of Books*, May 18, 2018. https://lareviewofbooks. org/article/the-ashtray-has-landed-the-case-of-morris-v-kuhn/. See also Joseph D. Martin, "A Filmmaker with an Ax to Grind Takes Aim at Thomas Kuhn's Legacy," review of *The Ashtray* by Errol Morris, *Science—Books,* May 22, 2018. https://blogs.sciencemag.org/ books/2018/05/22/the-ashtray/.
5. Martin, "A Filmmaker with an Ax to Grind."
6. Morris, *The Ashtray*, 28.
7. For an excellent essay that situates Kuhn within philosophical tradition and a broader academic context, see Philip Kitcher, "After Kuhn," in *The Oxford Handbook of Philosophy of Science,* ed. Paul Humphreys, (Oxford: Oxford University Press, 2016), 633–651. For a different perspective of Kuhn, but equally damning of Morris's caricature, see Ian Hacking, "Introductory Essay," in Thomas S. Kuhn, *The Structure of Scientific Revolutions*, 4th ed. (Chicago: University of Chicago Press, 2012).
8. Freeman Dyson recounts an encounter with Kuhn at a scientific meeting toward the end of his life in which Kuhn loudly and angrily made this proclamation for everyone to hear. See Freeman Dyson, *The Sun, the Genome, and the Internet: Tools of Scientific Revolution* (Oxford: Oxford University Press, 1999), 16.
9. Thomas Nickles, "Kuhn, Historical Philosophy of Science, and Case-Based Reasoning," *Configurations* 6, no. 1 (Winter 1998): 51–85. https:// muse.jhu.edu/article/8135.
10. Although there is ambiguity in Kuhn's use of paradigm throughout *SSR*, the real problems with his slippery language are usually traced to the final sections, where he explores what he means by scientific revolutions and their implications. See Kuhn, *The Structure of Scientific Revolutions*, Sections ix–xiii.
11. Kuhn, *The Structure of Scientific Revolutions*, Preface–xliii.
12. In particular, Kuhn detested his perceived association with relativists and social constructivists. For example, although sociologists of scientific knowledge (SSK) used *SSR* as a springboard to the founding

of their strong program, Kuhn made clear he did not subscribe to these extensions and projections of his work. Morris himself quotes Kuhn's clear rejection of SSK, but then immediately criticizes him for waffling! See Morris, *The Ashtray*, 30. For a more rigorous treatment of this, see Kitcher, "After Kuhn," 641–649.

13. Kuhn, *The Structure of Scientific Revolutions*, 1.
14. Otto, *The War On Science* and Morris, *The Ashtray*.
15. Oreskes, *Why Trust Science?*, 248. Oreskes is a distinguished historian of science, geologist, and science studies scholar. Her groundbreaking research, with Erik Conway, uncovered the vast coordinated misinformation campaigns waged by corporations and conservative think tanks. With the aid of a handful of well-placed scientists, these campaigns worked to seed doubt as to whether scientific consensus exists on issues like the anthropomorphic causes of climate change. See Oreskes and Conway, *Merchants of Doubt*.
16. Oreskes, *Why Trust Science?*, 19.
17. Oreskes, *Why Trust Science?*, 43–49.
18. Oreskes, *Why Trust Science?*, 54.
19. Oreskes, *Why Trust Science?*, 54.
20. Oreskes, *Why Trust Science?*, 50–59. This quote is from page 59, but the preceding section on feminist philosophers of science is important for context.
21. Oreskes, *Why Trust Science?*, 50–54.
22. Oreskes, *Why Trust Science?*, 51. Emphasis in the original.
23. Oreskes, *Why Trust Science?*, 52.
24. Oreskes, *Why Trust Science?*, 57.
25. Keating, "How My Nobel Dream Bit the Dust." This short memoir of the BICEP2 drama was adapted from his book *Losing the Nobel Prize*. Interestingly enough, the book has moments of myth-history, especially in the anecdotes Keating tells about the history of astronomy more broadly.
26. The BICEP2 collaboration was represented by members of its leadership team, including John M. Kovac (Harvard-Smithsonian CfA), Chao-Lin Kuo (Stanford/SLAC), Jamie Bock (Caltech/JPL), and Clem Pryke (University of Minnesota). In addition, Marc Kamionkowski (Johns Hopkins University) gave an outsider perspective and context for the announcement. BICEP2 Collaboration, "First Direct Evidence of Cosmic Inflation," press conference, Harvard-Smithsonian Center for Astrophysics (CfA), video 1:00:35,

Cambridge, MA, March 17, 2014. https://www.youtube.com/watch?reload=9&reload=9&v=Iasqtm1prlI.

27. Press release associated with the press conference; BICEP2 Collaboration and the CfA, "First Direct Evidence of Cosmic Inflation," press release no. 2014-05, Harvard-Smithsonian CfA, March 17, 2014. https://www.cfa.harvard.edu/news/2014-05.

28. BICEP2 and CfA, "First Direct Evidence of Cosmic Inflation."

29. During the press conference, Chao-Lin Kuo used this idea of inflation being the "bang" in Big Bang. See BICEP2 and CfA, "First Direct Evidence of Cosmic Inflation," press conference video, 10:20.

30. Dan Vergano, "Big Bang's 'Smoking Gun' Confirms Early Universe's Exponential Growth," *National Geographic Daily News*, March 17, 2014. http://news.nationalgeographic.com/news/2014/14/140317-big-bang-gravitational-waves-inflation-science-space/#.UymgsYXDWRg.

31. Kamionkowski made these remarks at 32:01. See BICEP2 and CfA, "First Direct Evidence of Cosmic Inflation," press conference video, 32:01.

32. The combined analysis of the Planck and BICEP/Keck teams can be found in Peter A.R. Ade, N. Aghanim, Z. Ahmed, R.W. Aikin, Kate Denham Alexander, M. Arnaud, J. Aumont, et al., "Joint Analysis of BICEP2/Keck Array and Planck Data," *Physical Review Letters* 114, no. 10 (2015): 101301. A clearly formulated review of this collaborative analysis can be found in Nils W. Halverson, "A Clearer View of a Dusty Sky," *Physics* 8 (2015): 21.

33. Peter Byrne, "A Bold Critic of the Big Bang's 'Smoking Gun,'" *Quanta Magazine*, July 24, 2014. https://www.scientificamerican.com/article/a-bold-critic-of-the-big-bang-s-smoking-gun/. This article includes an interview with David Spergel, Princeton University astrophysicist. A similar response to Spergel was attributed to Alan Guth. See Joel Achenbach, "BICEP2 Experiment's Big Bang Controversy Highlights Challenges for Modern Science," *The Washington Post, Health and Science*, July 23, 2014. https://www.washingtonpost.com/national/health-science/bicep2-experiments-big-bang-controversy-highlights-challenges-for-modern-science/2014/07/23/707bc9e6-02c6-11e4-b8ff-89afd3fad6bd_story.html.

34. Keating, "How My Nobel Dream Bit the Dust."

35. See Shapin, "Why Scientists Shouldn't Write History."

36. The following analysis relies on the published transcript of Caputo's podcast "Learning Curve," in which he interviewed NIAID Director, and Coronavirus Task Force member Anthony Fauci. Anthony

Fauci, "Science Is Truth," interview by Michael Caputo, *Learning Curve*, HHS.gov Podcasts, June 17, 2020, audio, 36:12. https://www.hhs.gov/podcasts/learning-curve/learning-curve-05-dr-anthony-fauci-science-is-truth.html.

37. Oreskes, *Why Trust Science?*, 51. Emphasis in the original.
38. Adichie, "The Danger of a Single Story," video: 17:32.

Bibliography

Abbot, B.P., et al. "Observation of Gravitational Waves from a Binary Black Hole Merger." *Physical Review Letters* 116, 061102 (2016): 1–16.

Accili, Di Neda. "Qui C'è Aria di Terremoto." *Club 3* (October 2008): 50–51. http://www.stpauls.it/club3/0810c3/1008Art_Inventore.pdf.

Achenbach, Joel. "BICEP2 Experiment's Big Bang Controversy Highlights Challenges for Modern Science." *The Washington Post, Health and Science*, July 23, 2014. https://www.washingtonpost.com/national/health-science/bicep2-experiments-big-bang-controversy-highlights-challenges-for-modern-science/2014/07/23/707bc9e6-02c6-11e4-b8ff-89afd3fad6bd_story.html.

Ackerman, Todd. "Chinese Scientist, Assisted by Rice Professor, Claims First Gene-Edited Babies." *AP News*, November 26, 2018. https://apnews.com/94fecf56cac841639d3bc9bad8db1698.

Ade, Peter A.R., N. Aghanim, Z. Ahmed, R.W. Aikin, Kate Denham Alexander, M. Arnaud, J. Aumont, et al. "Joint Analysis of BICEP2/Keck Array and Planck Data." *Physical Review Letters* 114, no. 10 (2015): 101301.

Adichie, Chimamanda Ngozi. "The Danger of a Single Story." Filmed July 2009 at TEDGlobal. TED video, 18:34. https://www.ted.com/talks/chimamanda_ngozi_adichie_the_danger_of_a_single_story?language=en#t-1114935.

Ahn, Janet N., et al. "Even Einstein Struggled: Effects of Learning About Great Scientists' Struggles on High School Students' Motivation to Learn Science." *Journal of Educational Psychology* 108, no. 3 (2016): 314–328.

Akst, Jef. "EPO Revokes Broad's CRISPR Patent." *The Scientist: Exploring Life, Inspiring Innovation*, January 17, 2018. https://www.the-scientist.com/the-nutshell/epo-revokes-broads-crispr-patent-30400.

Alder, Ken. "The History of Science as Oxymoron: From Scientific Exceptionalism to Episcience." *Isis* 104, no. 1 (March 2013): 88–101.

Alexander, David E. "The L'Aquila Earthquake of 6 April 2009 and Italian Government Policy on Disaster Response." *Journal of Natural Resources Policy Research* 2, no. 4 (October 2010): 325–342.

Alexander, D.E. "Communicating Earthquake Risk to the Public: The Trial of the 'L'Aquila Seven'." *Journal of the International Society for the Prevention and Mitigation of Natural Hazards* 72, no. 2 (2014): 1159–1173.

Alvargonzález, David. "Is the History of Science Essentially Whiggish?" *History of Science* 51, no. 1 (March 2013): 85–99.

Amrhein, Valentin, David Trafimow, and Sander Greenland. "Inferential Statistics as Descriptive Statistics: There Is No Replication Crisis if We Don't Expect Replication." *American Statistician* 73, Supplement 1 (2019): 262–270.

Anderson, Benedict. *Imagined Communities: Reflections on the Origin and Spread of Nationalism*, Reprint ed. London: Verso, 2016.

Assmann, Jan. *Religion and Cultural Memory*. Translated by Rodney Livingstone. Stanford: Stanford University Press, 2006.

Bacciagaluppi, Guido and Antony Valentini, eds. *Quantum Theory at the Crossroads: Reconsidering the 1927 Solvay Conference*. Cambridge: Cambridge University Press, 2009.

Baker, Monya. "1,500 Scientists Lift the Lid on Reproducibility." *Nature News* 533, no. 7604 (2016): 452–454.

Barthes, Roland. *Mythologies*. (1952.) Translated by Anette Lavers. New York: Hill and Wang, 1972.

Bartusiak, Marcia. *Einstein's Unfinished Symphony: The Story of a Gamble, Two Black Holes, and a New Age of Astronomy*. New Haven: Yale University Press, 2017.

Beede, David, et al. "Women in STEM: A Gender Gap to Innovation," U.S. Department of Commerce, Economics and Statistics Administration, ESA Issue Brief #04-11 August 2011, 1.

Bell, Duncan S.A. "Mythscapes: Memory, Mythology, and National Identity." *The British Journal of Sociology* 54, no. 1 (2003): 63–81.

Bell, John S. "On the Impossible Pilot Wave." *Foundations of Physics* 12, no. 10 (October 1982): 989–999. As reproduced in John S. Bell, *Speakable and Unspeakable in Quantum Mechanics*. Cambridge: Cambridge University Press, 1987.

Beller, Mara. *Quantum Dialogue: The Making of a Revolution*. Chicago: The University of Chicago Press, 2001.

Belousek, Darren. "Einstein's 1927 Unpublished Hidden Variables Theory: Its Background, Context and Significance." *Studies in History and Philosophy of Modern Physics* 21, no. 4 (1996): 437–461.

Benessia, Alice and Bruna De Marchi, "When the Earth Shakes . . . and Science with It. The Management and Communication of Uncertainty in the L'Aquila Earthquake." *Futures* 91 (2017): 35–45.

Bennetts, Leslie. "A Master of Mythology Is Honored." *The New York Times*, March 1, 1985. http://www.nytimes.com/1985/03/01/books/a-master-of-mythology-is-honored.html.

Bera, Rajendra K. "The Story of the Cohen-Boyer Patents." *Current Science* 96, no. 6 (March 25, 2009): 760–763.

Bertolotti. Mario. *The History of the Laser*. London: Institute of Physics, 2005.

BICEP2 Collaboration and the CfA. "First Direct Evidence of Cosmic Inflation." Press conference, Harvard-Smithsonian Center for Astrophysics (CfA), video 1:00:35. Cambridge, MA, March 17, 2014. https://www.youtube.com/watch?reload=9&reload=9&v=Iasqtm1prlI.

BICEP2 Collaboration and the CfA. "First Direct Evidence of Cosmic Inflation." Press release no. 2014-05. The Harvard-Smithsonian Center for Astrophysics (CfA), March 17, 2014. https://www.cfa.harvard.edu/news/2014-05.

Billi, Marco (Judge). Tribunale di L'Aquila, Sezione Penale. Motivazione Sentenza n. 380 del 22/10/2012, Depositata il 19/01/2013. http://www.magistraturademocratica.it/mdem/qg/doc/Tribunale_di_LAquila_sente nza_condanna_Grandi_Rischi_terremoto.pdf.

Bird, Kai and Martin J. Sherwin. *American Prometheus: The Triumph and Tragedy of J. Robert Oppenheimer*. New York: Knopf Publishing, 2005.

Blum, Alexander, Roberto Lalli, and Jürgen Renn. "The Reinvention of General Relativity." *Isis* 106, no. 3 (September 2015): 598–620.

Bohm, David. *Quantum Theory*. New Jersey: Prentice-Hall, 1951.

Bohm, David. "A Suggested Interpretation of Quantum Theory in Terms of 'Hidden' Variables. I." *Physical Review* 85, no. 2 (January 1952): 166–179.

Bohm, David. "A Suggested Interpretation of Quantum Theory in Terms of 'Hidden' Variables. II." *Physical Review* 85, no. 2 (January 1952): 180–193.

Bohm, David. FBI FOIA File #100-207045.

Bohm, David. Interview with Maurice Wilkins (1986–1987). Sixteen tapes and unedited transcripts are part of Bohm's papers at Birkbeck College Library, London and American Institute of Physics Papers.

Bolinska, Agnes and Joseph D. Martin, "Negotiating History: Contingency, Canonicity, and Case Studies." *Studies in History and Philosophy of Science Part A* 80 (2020): 37–46.

Bolt, Bruce A. "Locating Earthquakes and Plate Boundaries." In *Plate Tectonics: An Insider's History of the Modern Theory of the Earth*, edited by Naomi Oreskes. London: CRC Press, 2018.

Bondi, Hermann. "Plane Gravitational Waves in General Relativity." *Nature* 179 (1957): 1072–1073.

Born, Max. "Hilbert und die Physik." *Die Naturwissenschaften* 10 (1922): 88–93.

Born, Max. *Natural Philosophy of Cause and Chance*. New York: Dover Publications, 1964.

Born, Max and Werner Heisenberg. "Quantum Mechanics." In the Fifth Solvay Council Proceedings, produced in 1928; reprinted in *Quantum Theory at the Crossroads: Reconsidering the 1927 Solvay Conference*, edited by Guido Bacciagaluppi and Antony Valentini, 398. Cambridge: Cambridge University Press, 2009.

Breakthrough Prize Foundation. "Recipients of the 2015 Breakthrough Prizes in Fundamental Physics and Life Sciences Announced." Press release, *Breakthroughprize.org*, November 9, 2014. https://breakthrough-prize.org/News/21.

Brillouin, Léon. "The New Atomic Mechanics." *Journal de Physique et le Radium* 7 (1926): 135. Reprinted in *Selected Papers on Wave Mechanics*. Translated by Winifred Deans. London: Blackie & Son Limited, 1928.

Burke, Peter. *History and Social Theory*. 2nd ed. Cornell: Cornell University Press, 2005.

Bush, John W.M. "The New Wave of Pilot-Wave Theory." *Physics Today* 68, no. 8 (August 2015): 47–53.

Butterfield, Herbert. *The Whig Interpretation of History*. London: Bell and Sons, 1931.

Byrne, Peter. "A Bold Critic of the Big Bang's 'Smoking Gun.'" *Quanta Magazine*, July 24, 2014. https://www.scientificamerican.com/article/a-bold-critic-of-the-big-bang-s-smoking-gun/.

Camilleri, Kristian. "Constructing the Myth of the Copenhagen Interpretation." *Perspectives on Science* 17, no. 1 (2009): 26–57.

Camilleri, Kristian. *Heisenberg and the Interpretation of Quantum Mechanics: The Physicist as Philosopher*. Cambridge: Cambridge University Press, 2009.

Campbell, Joseph with Bill Moyers. *The Power of Myth*. New York: Anchor Books, 1991.

Canales, Jimena. *The Physicist and the Philosopher: Einstein, Bergson, and the Debate that Changed our Understanding of Time*. Princeton: Princeton University Press, 2015.

Cartlidge, Edwin. "Updated: Appeals Court Overturns Manslaughter Convictions of Six Earthquake Scientists." *ScienceMag News*, November 10, 2014. https://www.sciencemag.org/news/2014/11/updated-appeals-court-overturns-manslaughter-convictions-six-earthquake-scientists.

Cartlidge, Edwin. "Italy's Supreme Court Clears L'Aquila Earthquake Scientists for Good." *ScienceMag News*, November 20, 2015. https://www.sciencemag.org/news/2015/11/italy-s-supreme-court-clears-l-aquila-earthquake-scientists-good.

Cartlidge, Edwin. "Seven-year Legal Saga Ends as Italian Official is Cleared of Manslaughter in Earthquake Trial." *ScienceMag News*, October 3, 2016. https://www.sciencemag.org/news/2016/10/seven-year-legal-saga-ends-italian-official-cleared-manslaughter-earth-quake-trial.

Cassidy, David C. *Beyond Uncertainty: Heisenberg, Quantum Physics, and the Bomb*. New York: Bellevue Literary Press, 2009.

Cassirer, Ernest. *Language and Myth*. New York: Dover, 1953.

Chartier, Roger. *Inscription & Erasure: Literature and Written Culture from the Eleventh to the Eighteenth Century*. Translated by Arthur Goldhammer. Philadelphia: University of Pennsylvania Press, 2007.

Cho, Adrian. "Here's the First Person to Spot Those Gravitational Waves." *ScienceMag*, February 11, 2016. https://www.sciencemag.org/news/2016/02/here-s-first-person-spot-those-gravitational-waves.

Cho, Adrian. "Remembering Joseph Weber, the Controversial Pioneer of Gravitational Waves." *ScienceMag*, February 12, 2016. https://www.sciencemag.org/news/2016/02/remembering-joseph-weber-controversial-pioneer-gravitational-waves.

Chow, Andrew R. "The Significance of Brienne's Tribute to Jaime in the Game of Thrones Finale." *Time Magazine*, May 20, 2019. https://time.com/5591842/game-of-thrones-brienne-finale/.

Cohen, Jon. "How the Battle Lines over CRISPR Were Drawn." *ScienceMag*, 2, February 15, 2017. http://www.sciencemag.org/news/2017/02/how-battle-lines-over-crispr-were-drawn.

Cohen, Jon. "Round One of CRISPR Patent Legal Battle Goes to the Broad Institute," *ScienceMag*, 2, February 15, 2017. http://www.sciencemag.org/news/2017/02/round-one-crispr-patent-legal-battle-goes-broad-institute.

Cohen, Jon. "Federal Appeals Court Hears CRISPR Patent Dispute." *ScienceMag*, 4, April 30, 2018. http://www.sciencemag.org/news/2018/04/federal-appeals-court-hears-crispr-patent-dispute.

Cohen, Jon. "With Prestigious Prize, an Overshadowed CRISPR Researcher Wins the Spotlight." *ScienceMag* 6, June 4, 2018. https://www.sciencemag.org/news/2018/06/prestigious-prize-overshadowed-crispr-researcher-wins-spotlight.

Collins, Harry M. "The Place of the 'Core-Set' in Modern Science: Social Contingency with Methodological Propriety in Science." *History of Science* 19, no. 1 (1981): 6–19.

Collins, Harry. *Gravity's Shadow: The Search for Gravitational Waves.* Chicago: The University of Chicago Press, 2004.

Collins, Harry and Trevor Pinch. *The Golem: What You Should Know About Science.* 2nd ed. Cambridge: Cambridge University Press, 2012.

Collins, Harry and Trevor Pinch. *The Golem at Large: What You Should Know About Technology.* Cambridge: Cambridge University Press, 2014.

Collins, Harry. *Gravity's Kiss: The Detection of Gravitational Waves.* Cambridge, MA: MIT Press, 2017.

Collins, Harry. *Forms of Life: The Method and Meaning of Sociology.* Cambridge, MA: The MIT Press, 2019.

Comfort, Nathanial. "A Whig History of CRISPR." *Genotopia: Here Lies Truth.* Published online January 18, 2016. https://genotopia.scienceblog. com/573/a-whig-history-of-crispr/.

Committee on Un-American Activities, House of Representatives, U.S. Congress. *Hearings Regarding Communist Infiltration of Radiation Laboratory and Atomic Bomb Project at the University of California, Berkeley, Calif.,* Vols. 2–3. Washington, DC: U.S. Government Printing Office, 1951.

Contreras, Jorge L. and Jacob S. Sherkow. "CRISPR, Surrogate Licensing, and Scientific Discovery." *Science* 355, no. 6326 (2017): 698–700.

Cushing, James T. *Quantum Mechanics: Historical Contingency and the Copenhagen Hegemony.* Chicago: The University of Chicago Press, 1994.

Cynober, Timothé. "CRISPR: One Patent to Rule Them All." *Labiotech,* February 11, 2019. https://www.labiotech.eu/in-depth/crispr-patent-dispute-licensing/.

Dacey, James. "A Radon Detector for Earthquake Prediction." *Physicsworld,* March 18, 2010. https://physicsworld.com/a/a-radon-detector-for-earthquake-prediction/.

Darrigol, Olivier. "The Genesis of the Theory of Relativity." In *Einstein, 1905–2005: Poincaré Seminar 2005,* edited by Thibault Damour, Olivier Darrigol, and Vincent Rivasseau. Berlin: Springer Science & Business Media, 2006.

Daston, Loraine. "Science Studies and the History of Science." *Critical Inquiry* 35, no. 4 (Summer, 2009): 798–813.

Daston, Lorraine and Peter Galison. *Objectivity.* New York: Zone Books, 2007.

De Bernardinis, Bernardo. "Curriculum Vitae di Bernardo De Bernardinis aggiornato agosto 2013." *ISPRA.* https://www.isprambiente.gov.it/

files/trasparenza/organizzazione/cda/CURRICULUM_DE_Bernardinis.
pdf/view.

De Broglie, Louis. "La Structure Atomique de la Matière et du
Rayonnement et la Mécanique Ondulatoire." *Comptes Rendus* 184
(1927): 273–274.

De Broglie, Louis. "La Mécanique Ondulatoire et la Structure Atomique
de la Matière et du Rayonnement." *Journal de Physique et du Radium* 8
(1927): 225–241.

De Broglie, Louis. *An Introduction to the Study of Wave Mechanics*. Translated by
H.T. Flint. London: Methuen & Co. Ltd., 1930.

De Broglie, Louis. *Matter and Light—The New Physics*. Translated by W.H.
Johnston. New York: Norton and Co., 1939.

De Vrieze, Jop. "Bruno Latour, a Veteran of the 'Science Wars,' Has a
New Mission." *Science*Insider, October 10, 2017. https://www.
sciencemag.org/news/2017/10/bruno-latour-veteran-science-wars-
has-new-mission.

Dear, Peter and Sheila Jasanoff. "Dismantling Boundaries in Science and
Technology Studies." *Isis* 101, no. 4 (December 2010): 759–774.

Del Fante, Dario. "In Margine al Terremoto de L'Aquila: Il Caso
Giuliani," Master's Thesis, University of Siena, 2013. https://www.
academia.edu/13446373/IN_MARGINE_AL_TERREMOTO_DE_L_A
QUILA_IL_CASO_GIULIANI.

DeVasto, Danielle, et al. "Stasis and Matters of Concern: The Conviction
of the L'Aquila Seven." *Journal of Business and Technical Communication* 30,
no. 2 (2016): 131–164.

Dieks, Dennis. "Von Neumann's Impossibility Proof: Mathematics in
the Service of Rhetorics." *Studies in History and Philosophy of Science Part B*
60 (November 2017): 136–148.

Dirac, P.A.M. "Quantum Mechanics of Many-Electron Systems."
Proceedings of the Royal Society London A123 (1929): 714–733.

Dollar, J. "The Man Who Predicted an Earthquake." *The Guardian*, April 5,
2010.   https://www.theguardian.com/world/2010/apr/05/laquila-
earthquake-prediction-giampaolo-giuliani.

Doudna, Jennifer A. and Samuel H. Sternberg. *A Crack in Creation: Gene
Editing and the Unthinkable Power to Control Evolution*. Boston: Houghton
Mifflin Harcourt, 2017.

Durkheim, Émile. *The Elementary Forms of Religious Life*. Translated by
Karen E. Fields. New York: Free Press, 1995.

Dyson, Freeman. *The Sun, the Genome, and the Internet: Tools of Scientific
Revolution*. Oxford: Oxford University Press, 1999.

Eckert, Michael. "The Emergence of Quantum Schools: Munich, Göttingen and Copenhagen as New Centers of Atomic Theory." *Annalen der Physik* 10 (2001): 151–162.

Eddington, A.S. "The Future of International Science." *Observatory* 39 (June 1916): 271.

Einstein, Albert. *Essays in Science*. Dover ed. New York: Dover, 2009.

Eisen, Michael. "The Villain of CRISPR." michaeleisen.org (blog). Published online January 25, 2016. http://www.michaeleisen.org/blog/?p=1825.

Eliade, Mircea. *The Myth of the Eternal Return: Cosmos and History*. Princeton: Princeton University Press, 1971.

Fauci, Anthony. "Science Is Truth." Interview by Michael Caputo. *Learning Curve*, HHS.gov Podcasts, June 17, 2020. Audio, 36:12. https://www.hhs.gov/podcasts/learning-curve/learning-curve-05-dr-anthony-fauci-science-is-truth.html.

Farland, Dennis. "The Effect of Historical, Nonfiction Trade Books on Elementary Students' Perceptions of Scientists." *Journal of Elementary Science Education* 18, no. 2 (Fall 2006): 31–48.

Feder, Toni. "Q&A: Virginia Trimble on 50-plus Years in Astronomy." *Physics Today*, March 13, 2018. https://physicstoday.scitation.org/do/10.1063/pt.6.4.20180313a/full/.

Fernández, Clara Rodríguez. "Ten Unusual Applications of CRISPR Gene Editing." *Labiotech*, March 4, 2019. https://www.labiotech.eu/tops/crispr-applications-gene-editing/.

Feynman, Richard. "The Development of the Space-Time View of Quantum Electrodynamics." *Nobel Prize Lecture*, December 11, 1965. https://www.nobelprize.org/prizes/physics/1965/feynman/lecture/.

Feynman, Richard. "What Is Science?" *The Physics Teacher* 7, no. 6 (September 1969): 313–320.

Feynman, Richard. "Take the World from Another Point of View." Yorkshire Public Television Program (1973). Video file, 36:41. YouTube. Posted by mrtp, May 28, 2015. https://www.youtube.com/watch?v=GNhlNSLQAFE.

Feynman, Richard P. and Ralph Leighton. *"Surely You're Joking, Mr. Feynman!": Adventures of a Curious Character*. New York: Random House, 1992.

Feynman, Richard. *QED: The Strange Theory of Light and Matter*, 3rd ed. Princeton: Princeton University Press, 2006.

Fine, Arthur. *The Shaky Game: Einstein Realism and the Quantum Theory*, 2nd ed. Chicago: The University of Chicago Press, 1996.

Forman, Paul. "Scientific Internationalism and the Weimar Physicists: The Ideology and Its Manipulation in Germany after World War I." *Isis* 64 (1973): 150–180.

Forstner, Christian. "The Early History of David Bohm's Quantum Mechanics through the Perspective of Ludwik Fleck's Thought Collectives." *Minerva* 46, no. 2 (2008): 215–229.

Foucault, Michel. "Nietzsche, Genealogy, History." In *Language, Counter-Memory, Practice: Selected Essays and Interviews*, edited by Donald F. Bouchard. Translated by Donald F. Bouchard and Sherry Simon, 139–164. Ithaca: Cornell University Press, 1977.

Franklin, Allan. "How to Avoid the Experimenters' Regress." *Studies in History and Philosophy of Science* 25, no. 3 (1994): 463–491.

Franklin, Allan. "Gravity Waves and Neutrinos: The Latter work of Joseph Weber." *Perspectives in Science* 18, no. 2 (2010): 119–151.

Franklin, Allan and Harry Collins. "Two Kinds of Case Study and a New Agreement." In *The Philosophy of Historical Case Studies*, Boston Studies in the Philosophy and History of Science, edited by Tilman Sauer and Raphael Scholl, 88–115. Dordrecht: Springer, 2016.

Freire Jr., Olival. "Quantum Controversy and Marxism," *Historia Scientiarum* 7, no. 2 (1997): 137–152.

Freire Jr., Olival. "Science and Exile: David Bohm, the Hot Times of the Cold War, and His Struggle for a New Interpretation of Quantum Mechanics." *Historical Studies in the Physical and Biological Sciences* 36, no. 1 (September 2005): 1–34.

Freire Jr., Olival. "Orthodoxy and Heterodoxy in the Research on the Foundations of Quantum Physics: E.P. Wigner's Case." In *Cognitive Justice in a Global World: Prudent Knowledges for a Decent Life*, edited by Boaventura de Sousa Santos. Langham: Lexington Books, 2007.

Freire Junior, Olival. *The Quantum Dissidents: Rebuilding the Foundations of Quantum Mechanics (1950–1990)*. Berlin: Springer, 2015.

Freire Junior, Olival. *David Bohm: A Life Dedicated to Understanding the Quantum World*. Switzerland: Springer, 2019.

Gaddis, John Lewis. *The Landscape of History: How Historians Map the Past*. New York: Oxford University Press, 2002.

Galison, Peter. *Einstein's Clocks, Poincaré's Maps: Empires of Time*. New York: W.W. Norton, 2004.

Galison, Peter. "Solvay Redivivus." In *Proceedings of the 23rd Solvay Conference on Physics: The Quantum Structure of Space and Time*, edited by David Gross et al., 1–18. Singapore: World Scientific Publishing, 2007.

Galison, Peter L., Gerald Holton, and Silvan S. Schweber, eds. *Einstein for the 21st Century: His Legacy in Science, Art, and Modern Culture*. Princeton: Princeton University Press, 2008.

Ganz, Cheryl R. *The 1933 Chicago World's Fair: A Century of Progress*. Urbana: University of Illinois Press, 2008.

Geertz, Clifford. "Deep Play: Notes on the Balinese Cockfight." *Dædalus Journal of the American Academy of Arts and Sciences* 101, no. 1 (Winter 1972): 1–38.

Geller, Robert J. "Earthquake Prediction: A Critical Review." *Geophysical Journal International* 131, no. 3 (1997): 425–450.

Genentech. "Genentech Announces Vice President Appointment in Research." Press release, January 21, 2009. https://www.fiercebiotech.com/biotech/genentech-announces-vice-president-appointment-research.

Gieryn, Thomas F. "Boundary-Work and the Demarcation of Science from Non-Science: Strains and Interests in Professional Ideologies of Scientists." *American Sociological Review* 48, no. 6 (1983): 781–795.

Gleick, James. *Genius: The Life and Science of Richard Feynman*. New York: Vintage Books, 1993.

Goldberg, Lesley. "Jennifer Lopez Sets Futuristic Bio-Terror Drama at NBC (Exclusive)." *The Hollywood Reporter*, October 18, 2016. https://www.hollywoodreporter.com/live-feed/jennifer-lopez-sets-futuristic-bio-939509.

Gordin, Michael. *The Pseudoscience Wars: Immanuel Velikovsky and the Birth of the Modern Fringe*. Chicago: The University of Chicago Press, 2013.

Grant, Rachel A. and Tim Halliday. "Predicting the Unpredictable; Evidence of Pre-Seismic Anticipatory Behaviour in the Common Toad." *Journal of Zoology* 281, no. 4 (2010): 263–271.

Grens, Kerry. "UC Berkeley Team to Be Awarded CRISPR Patent." *The Scientist: Exploring Life, Inspiring Innovation*, February 11, 2019. https://www.the-scientist.com/news-opinion/uc-berkeley-team-to-be-awarded-crispr-patent-65453.

Gribbin, John. *In Search of Schrödinger's Cat: Quantum Physics and Reality*. New York: Bantam Books, 1984.

Griffiths, David J. *Revolutions in Twentieth-Century Physics*. Cambridge: Cambridge University Press, 2013.

Gross, Alan G. "Do Disputes over Priority Tell Us Anything about Science?" *Science in Context* 11, no. 2 (Summer 1998): 161–179.

Grossman, Lionel. "Anecdote and History." *History and Theory* 42 (May 2003): 143–168.

Gupta, R.M. and K. Musunuru. "Expanding the Genetic Editing Tool Kit: ZFNs, TALENs, and CRISPR-Cas9." *Journal of Clinical Investigation* 124, no. 10 (October 2014): 4154–4161.

Hacking, Ian. "Introductory Essay." In Thomas S. Kuhn, *The Structure of Scientific Revolutions*, 4th ed. Chicago: University of Chicago Press, 2012.

Hall, Rupert A. "On Whiggism." *History of Science* 21, no. 1 (1983): 45–59.

Hall, Stephen S. "Scientists on Trial: At Fault?" *Nature News*, September 14, 2011. https://www.nature.com/news/2011/110914/full/477264a.html.

Halpern, Paul. *Einstein's Dice and Schrodinger's Cat: How Two Great Minds Battled Quantum Randomness to Create a Unified Theory of Physics*. Philadelphia: Basic Books, 2015.

Halverson, Nils W. "A Clearer View of a Dusty Sky." *Physics* 8 (2015): 21.

Harper, Kristine. *Weather by the Numbers: The Genesis of Modern Meteorology.* Cambridge, MA: The MIT Press, 2008.

Harrison, Edward. "Whigs, Prigs and Historians of Science." *Nature* 329, no. 6136 (September 17, 1987): 213–214.

Heilbron, John. "The Earliest Missionaries of the Copenhagen Spirit." *Revue d'histoire des sciences* 38, no. 3 (1985): 195–230.

Heisenberg, Werner. *The Physical Principles of the Quantum Theory.* Chicago: The University of Chicago Press, 1930.

Hill, Jonathan D., ed. Rethinking History and Myth: Indigenous South American Perspectives on the Past. Urbana: University of Illinois Press, 1988.

History of Science Society (HSS). "History of the Society." About HSS. https://hssonline.org/about/history-of-the-society/.

Hobson, Art. *Tales of the Quantum: Understanding Physics' Most Fundamental Theory.* New York: Oxford University Press, 2017.

Holton, Gerald J. *Science and Anti-science.* Cambridge, MA: Harvard University Press, 1993.

Holton, Gerald. "Who Was Einstein? Why Is He Still So Alive?" In *Einstein for the 21st Century: His Legacy in Science, Art, and Modern Culture*, edited by Peter L. Galison, Gerald Holton, and Silvan S. Schweber. Princeton: Princeton University Press, 2008.

Hong, Huang-Yao and Xiaodong Lin-Siegler. "How Learning about Scientists' Struggles Influences Students' Interest and Learning in Physics." *Journal of Educational Psychology* 104 (2012): 469–484.

Hough, Susan E. "Earthquakes: Predicting the Unpredictable?" *Geotimes* 50, no. 3 (March 2005). http://www.geotimes.org/mar05/feature_eqprediction.html.

Hough, Susan E. *Predicting the Unpredictable: The Tumultuous Science of Earthquake Prediction*. Princeton: Princeton University Press, 2016.

Houldsworth, Adam. "Who Owns the Most CRISPR Patents Worldwide? Surprisingly, It's Agrochemical Giant DowDuPont." *Genetic Literacy Project: Science not Ideology*, February 16, 2018. https://geneticliteracyproject.org/2018/02/16/owns-crispr-patents-worldwide-surprisingly-agrochemical-giant-dowdupont/.

Howard, Don. "Who Invented the 'Copenhagen Interpretation'? A Study in Mythology." *Philosophy of Science* 71, no. 5 (December 2004): 669–682.

Howells, Richard. *The Myth of the Titanic*, Centenary ed. New York: Palgrave Macmillan, 2012.

Hull, David L. "In Defense of Presentism." *History and Theory* 18, no. 1 (February 1979): 1–15.

Igarashi, G., et al. "Ground-Water Radon Anomaly before the Kobe Earthquake in Japan." *Science* 269, no. 5220 (1995): 60–61.

Istituto Nazionale di Geofisica e Vulcanologia (INGV). "Lista Terremoti." INGV.it. http://terremoti.ingv.it.

Istituto Nazionale di Fisica Nucleare (INFN). "Laboratori Nazionali del Gran Sasso (LNGS)—LVD." INFN.it. https://www.lngs.infn.it/en/lvd.

Istituto Nazionale di Fisica Nucleare (INFN). "Laboratori Nazionali del Gran Sasso (LNGS)—ERMES." INFN.it. https://www.lngs.infn.it/en/ermes.

Institute for Advanced Study (IAS). "Thomas S. Kuhn." Scholars. https://www.ias.edu/scholars/thomas-s-kuhn.

Isaacson, Walter. *Einstein: His Life and Universe*. New York: Simon & Schuster, 2007.

Ito, Kenji. "The *Geist* in the Institute: The Production of Quantum Physics in 1930s Japan." In *Pedagogy and the Practice of Science*, edited by David Kaiser, 151–183. Cambridge, MA: The MIT Press, 2005.

Jacobsen, Anja Skaar. "Léon Rosenfeld's Marxist Defense of Complementarity." *Historical Studies in the Physical and Biological Sciences* 37, no. Supplement (2007): 3–34.

Jackson, Myles W. *The Genealogy of a Gene: Patents, HIV/AIDS, and Race*. Cambridge, MA: MIT Press, 2015.

Jammer, Max. *The Conceptual Development of Quantum Mechanics (History of Modern Physics, 1800–1950)*. New York: McGraw-Hill, 1966.

Jammer, Max. *The Philosophy of Quantum Mechanics: The Interpretations of Quantum Mechanics in Historical Perspective*. New York: John Wiley & Sons, 1974.

Janssen, Michel and Christoph Lehner, eds. *The Cambridge Companion to Einstein*, Vol. 1. Cambridge: Cambridge University Press, 2014.

Jardine, Nick. "Whigs and Stories: Herbert Butterfield and the Historiography of Science." *History of Science* 41, no. 2 (June 1, 2003): 125–140.

Jasanoff, Sheila. "Future Imperfect: Science, Technology and the Imaginations of Modernity." In *Dreamscapes of Modernity: Sociotechnical Imaginaries and the Fabrication of Power*, edited by Sheila Jasanoff and Sang-Hyun Kim. Chicago: The University of Chicago Press, 2015.

Jasanoff, Sheila and Sang-Hyun Kim, eds. *Dreamscapes of Modernity: Sociotechnical Imaginaries and the Fabrication of Power*. Chicago: The University of Chicago Press, 2015.

Jinek, Martin, Krzysztof Chylinski, Ines Fonfara, Michael Hauer, Jennifer A. Doudna, and Emmanuelle Charpentier. "A Programmable Dual-RNA–Guided DNA Endonuclease in Adaptive Bacterial Immunity." *Science* 337, no. 6096 (2012): 816–821.

Joffe, Helene et al. "Stigma in Science: The Case of Earthquake Prediction." *Disasters* 42, no. 1 (2018): 81–100.

Johnson, Paul. "The Hidden Seismic Symphony in Earthquake Signals." *Scientific American175*, April 25, 2019. https://blogs.scientificamerican.com/observations/the-hidden-seismic-symphony-in-earthquake-signals/.

Jordan, T.H. et al. "Operational Earthquake Forecasting: State of Knowledge and Guidelines for Implementation." Final Report of the International Commission on Earthquake Forecasting for Civil Protection. *Annals of Geophysics* 54, no. 4 (2011): 315–391.

Jung, C.G. and Robert A. Segal, eds. *Jung on Mythology*, 2nd ed. Princeton: Princeton University Press, 1998.

Kaempffert, Waldemar. "Science in 151 Words." *The New York Times*, June 4, 1933.

Kaiser, David. "Making Tools Travel: Pedagogy and the Transfer of Skills in Post War Theoretical Physics." In *Pedagogy and the Practice of Science*, edited by David Kaiser, 41–74. Cambridge, MA: The MIT Press, 2005.

Kaiser, David. "Turning Physicists into Quantum Mechanics." *Physics World* (May 2007): 28–33.

Kaiser, David. *How the Hippies Saved Physics: Science, Counterculture, and the Quantum Revival*. New York: W.W. Norton & Company, 2011.

Kaiser, David. *Quantum Legacies: Dispatches from an Uncertain World*. Chicago: The University of Chicago Press, 2020.

Kanny, M. Allison, Linda J. Sax, and Tiffani A. Riggers-Piehl. "Investigating Forty Years of STEM Research: How Explanations for the Gender Gap Have Evolved over Time." *Journal of Women and Minorities in Science and Engineering* 20, 2 (2014): 127–148.

Kavli Prize. "2018 Kavli Prize in Nanoscience: A Conversation with Jennifer Doudna, Emmanuelle Charpentier and Virginijus Šikšnys." *Nanoscience*, November 9, 2018. http://kavliprize.org/events-and-features/2018-kavli-prize-nanoscience-conversation-jennifer-doudna-emmanuelle-charpentier.

Keating, Brian. *Losing the Nobel Prize: A Story of Cosmology, Ambition, and the Perils of Science's Highest Honor.* New York: W.W. Norton & Co., 2018.

Keating, Brian. "How My Nobel Dream Bit the Dust." *Nautilus*, Connections no. 059, April 19, 2018. http://nautil.us/issue/59/connections/how-my-nobel-dream-bit-the-dust.

Kennefick, Daniel. *Traveling at the Speed of Thought: Einstein and the Quest for Gravitational Waves.* Princeton: Princeton University Press, 2007.

Kerr, Richard A. "After the Quake, in Search of the Science—or Even a Good Prediction." *Science* 324 (April 2009): 322.

Kevles, Daniel J. *In the Name of Eugenics: Genetics and the Uses of Human Heredity.* New York: Knopf Doubleday Publishing Group, 1985.

Kevles, Daniel J. "Good-bye to the SSC: On the Life and Death of the Superconducting Super Collider." *Engineering and Science* 58, no. 2 (1995): 16–25.

Kevles, Daniel. *The Physicists: The History of a Scientific Community in Modern America.* Cambridge, MA: Harvard University Press, 1995.

Kiefer, Claus. "Space and Time 62 Years after the Berne Conference." Preprint. Accessed February 18, 2020: https://arxiv.org/pdf/1812.10679.pdf.

Kington, T. "L'aquila's Earthquake-Scarred Streets See Battle between Science and Politics." *The Guardian*, October 27, 2012. https://www.theguardian.com/world/2012/oct/27/laquila-earthquake-battle-science-politics.

Kinzel, Katherina. "Pluralism in Historiography: A Case Study of Case Studies." In *The Philosophy of Historical Case Studies*, Boston Studies in the Philosophy and History of Science, edited by Tilman Sauer and Raphael Scholl, 123–150. Dordrecht: Springer, 2016.

Kitcher, Philip. "After Kuhn." In *The Oxford Handbook of Philosophy of Science*, edited by Paul Humphreys, 633–651. Oxford: Oxford University Press, 2016.

Kitcher, Philip. "The Ashtray Has Landed: The Case of Morris v. Kuhn." Review of *The Ashtray*, by Errol Morris. *Los Angeles Review of Books*, May 18, 2018. https://lareviewofbooks.org/article/the-ashtray-has-landed-the-case-of-morris-v-kuhn/.

Klein, O. "The Atomicity of Electricity as a Quantum Theory Law." *Nature* 118, no. 2971 (1926): 516.

Klug, Aaron. "Rosalind Franklin and the Discovery of the Structure of DNA." *Nature* 219, no. 5156 (1968): 808.

Kojevnikov, Alexei. "David Bohm and Collective Movements." *Historical Studies in the Physical and Biological Sciences* 33, no. 1 (2002): 161–192.

Kojevnikov, Alexei. "Philosophical Rhetoric in Early Quantum Mechanics 1925–27: High Principles, Cultural Values and Professional Anxieties." In *Weimar Culture and Quantum Mechanics: Selected Papers by Paul Forman and Contemporary Perspectives on the Forman Thesis*, edited by Cathryn Carson, Alexei Kojevnikov, and Helmuth Trischler, 319–348. London: Imperial College Press, 2011.

Kolbert, Elizabeth. "The Shaky Science behind Predicting Earthquakes." *Smithsonian Magazine*, June 2015. https://www.smithsonianmag.com/science-nature/shaky-science-behind-predicting-earthquakes-180955296/.

Kozubek, James. *Modern Prometheus: Editing the Human Genome with CRISPR-Cas9*. Cambridge: Cambridge University Press, 2018.

Kragh, Helge. *Quantum Generations: A History of Physics in the Twentieth Century*. Princeton: Princeton University Press, 2002.

Krattiger, Anatole, et al. *Intellectual Property Management in Health and Agricultural Innovation: A Handbook of Best Practices*, Vol. 1. Oxford, UK: MIHR, and Davis, CA: PIPRA, 2007.

Kuhn, Thomas S. "The Essential Tension: Tradition and Innovation in Scientific Research?" In *The Essential Tension: Selected Studies in Scientific Tradition and Change*, edited by Thomas S. Kuhn, 225–239. Chicago: The University of Chicago Press, 1979.

Kuhn, Thomas S. *The Structure of Scientific Revolutions*, 4th ed. Chicago: The University of Chicago Press, 2012.

Kurosawa, Akira. *Rashomon* [film]. Producer: Daiei, Japan. Script: T. Matsuama (1950).

Lam, Barry. "The Ashes of Truth." *Hi-Phi Nation*. S1, episode 9. Podcast audio, April 18, 2017. https://hiphination.org/complete-season-one-episodes/episode-9-the-ashes-of-truth-april-18-2017/.

Lander, Eric S. "The Heroes of CRISPR." *Cell* 164, no. 1–2 (January 14, 2016): 18–28.

Latour, Bruno. *Science in Action: How to Follow Scientists and Engineers Through Society*. Cambridge, MA: Harvard University Press, 1987.

Leane, Elizabeth. *Reading Popular Physics: Disciplinary Skirmishes and Textual Strategies*. Farnham, UK: Ashgate Publishing, Ltd., 2007.

Lederman, Leon with Dick Teresi, *The God Particle: If the Universe Is the Answer, What Is the Question?* 2nd ed. New York: Mariner Books, 2006.

Ledford, Heidi. "How the US CRISPR Patent Probe Will Play Out." *Nature News* 531, no. 7593 (March 2016): 149. https://www.nature.com/news/how-the-us-crispr-patent-probe-will-play-out-1.19519.

Ledford, Heidi. "Titanic Clash over CRISPR Patents Turns Ugly." *Nature News* 537, no. 7621 (2016): 460. https://www.nature.com/news/titanic-clash-over-crispr-patents-turns-ugly-1.20631.

Leshner, Alan I. "Letter to the President of the Italian Republic." *American Association for the Advancement of Science (AAAS)*, June 30, 2010. http://www.aaas.org/sites/default/files/migrate/uploads/0630italy_letter.pdf.

Levenson, Thomas. *The Hunt for Vulcan: How Albert Einstein Destroyed a Planet, Discovered Relativity, and Deciphered the Universe*. New York: Random House, 2015.

Lévi-Strauss, Claude. "The Structural Study of Myth," *Journal of American Folklore* 68, no. 270 (1955): 428–444.

Lévi-Strauss, Claude. *Myth and Meaning*. New York: Schocken Books, 1978.

Levy, Richard S. "Political Antisemitism in Germany and Austria 1848–1914." In *Antisemitism: A History*, edited by Albert S. Lindemann and Richard S. Levy. Oxford: Oxford University Press, 2010.

LIGO Caltech. "What Is an Interferometer." *LIGO Caltech—Learn More*. https://www.ligo.caltech.edu/page/what-is-interferometer.

LIGO Caltech. "Gravitational Waves Detected 100 Years after Einstein's Prediction." Press release, February 11, 2016. https://www.ligo.caltech.edu/news/ligo20160211.

LIGO Scientific Collaboration (LSC). "Calibration of the Advanced LIGO Detectors for the Discovery of the Binary Black-Hole Merger GW150914." LIGO.org. https://www.ligo.org/science/Publication-GW150914Calibration/index.php.

Lodge, Oliver. "Truth or Convenience." *Nature* 119, no. 2994 (1927), 424.

Loison, Laurent. "Forms of Presentism in the History of Science. Rethinking the Project of Historical Epistemology," *Studies in History and Philosophy of Science* 60 (December 2016): 29–37.

Lorentz, Hendrik A. "General Discussion of the New Ideas Presented." In the Fifth Solvay Council Proceedings, produced in 1928; reprinted in *Quantum Theory at the Crossroads: Reconsidering the 1927 Solvay Conference*, edited by Guido Bacciagaluppi and Antony Valentini, 432–433. Cambridge: Cambridge University Press, 2009.

Maddox, Brenda. *Rosalind Franklin: The Dark Lady of DNA*. New York: Harper Perennial, 2003.

Majer, Ulrich. "The Axiomatic Method and the Foundations of Science: Historical Roots of Mathematical Physics in Göttingen (1900-1930)." In *John von Neumann and the Foundations of Quantum Physics*, edited by M. Rédei and M. Stöltzner. Dordrecht: Kluwer Academic Publishers, 2001.

Malinowski, Bronislaw. *Myth in Primitive Psychology*. New York: Read Books Ltd, 2014.

Martin, Joseph D. "A Filmmaker with an Ax to Grind Takes Aim at Thomas Kuhn's Legacy." Review of *The Ashtray* by Errol Morris. *Science—Books*, May 22, 2018. https://blogs.sciencemag.org/books/2018/05/22/the-ashtray/.

Mayr, Ernst. "When is Historiography Whiggish?" *Journal of the History of Ideas* 51, no. 2 (1990): 301–309.

Maza, Sarah. *The Myth of the French Bourgeoisie: An Essay on the Social Imaginary*. Cambridge, MA: Harvard University Press, 2005.

McNeill, William H. "Mythistory, or Truth, Myth, History, and Historians." *The American Historical Review*, Supplement to 91, no. 1. (February 1986): 1–10.

Mehra, Jagdish. *The Solvay Conferences on Physics: Aspects of the Development of Physics since 1911*. Boston: D. Reidel Publishing Co., 1975.

Mermin, N. David. "Could Feynman Have Said This?" *Physics Today* 57, no. 5 (2004): 10–11.

Merton, Robert K. "Science and Technology in a Democratic Order." *Journal of Legal and Political Sociology* 1, no. 1 (1942): 115–126.

Merton, Robert K. "Priorities in Scientific Discoveries: A Chapter in the Sociology of Science," *American Sociological Review* 22 (1957): 635–659.

Merton, Robert K. *The Sociology of Science: Theoretical and Empirical Investigations*. Chicago: University of Chicago Press, 1973.

Merton, Robert K. *On the Shoulders of Giants: A Shandean Postscript*. Chicago: University of Chicago Press, 1993.

Michaels, David. *The Triumph of Doubt: Dark Money and the Science of Deception*. New York: Oxford University Press, 2020.

Mika, Aggie. "Flux and Uncertainty in the CRISPR Patent Landscape." *The Scientist: Exploring Life, Inspiring Innovation*, October 2017. https://

www.the-scientist.com/bio-business/flux-and-uncertainty-in-the-crispr-patent-landscape-30228.

Mitroff, Ian I. "Norms and Counter-Norms in a Select Group of the Apollo Moon Scientists: A Case Study of the Ambivalence of Scientists." *American Sociological Review* 39, no. 4 (August 1974): 579–595.

Molho, Anthony and Gordon S. Wood, eds. *Imagined Histories: American Historians Interpret the Past.* Princeton: Princeton University Press, 1998.

Montreal *Gazette.* "Sterilization of Unfit Advocated: Feeble-Minded Increasing at Disproportionate Rate in Canada." *The Montreal Gazette,* November 14, 1933. https://news.google.com/newspapers?id=oBYvA AAAIBAJ&pg=6315,1778929.

Moro-Abadía, Oscar. "Thinking About 'Presentism' From a Historian's Perspective: Herbert Butterfield and Hélène Metzger." *History of Science* 47, no. 1 (March 2009): 55–77.

Morris, Errol. *The Ashtray (Or the Man Who Denied Reality).* Chicago: The University of Chicago Press, 2018.

Munafò, M., Nosek, B., Bishop, D., et al. "A Manifesto for Reproducible Science." *Nature Human Behaviour* 1, no. 0021 (2017): 1–9. https://www.nature.com/articles/s41562-016-0021.

National Science Foundation (NSF). "LIGO Detects Gravitational Waves—Announcement at Press Conference." *NSF Press Conference,* video 1:11:29, February 11, 2016. https://www.ligo.caltech.edu/video/ligo20160211v11.

National Institute of Health. "What Are Genome Editing and CRISPR-Cas9?" https://ghr.nlm.nih.gov/primer/genomicresearch/genomeediting.

Newton Project. "About the Newton Project." http://www.newtonproject.ox.ac.uk/about-us/newton-project.

Neffe, Jürgen. *Einstein: A Biography.* Translated by Shelley Frisch. New York: Farrar, Straus and Giroux, 2007.

Nickles, Thomas. "Kuhn, Historical Philosophy of Science, and Case-Based Reasoning." *Configurations* 6, no. 1 (Winter 1998): 51–85. https://muse.jhu.edu/article/8135.

NobelPrize.org. "The Nobel Prize in Physics 1965." Nobel Prizes & Laureates. https://www.nobelprize.org/prizes/physics/1965/summary/.

Novick, Peter. *That Nobel Dream: The "Objectivity Question" and the American Historical Profession.* Cambridge: Cambridge University Press, 1988.

Numbers, Ronald L., ed. *Galileo Goes to Jail: And Other Myths About Science and Religion.* Cambridge, MA: Harvard University Press, 2009.

Numbers, Ronald L. and Kostas Kampourakis, eds. *Newton's Apple and Other Myths About Science*. Cambridge, MA: Harvard University Press, 2015.

Nye, Mary J. "Aristocratic Culture and the Pursuit of Science: The De Broglies in Modern France." *Isis* 88, no. 3 (September 1997): 397–421.

Odin. "DIY Bacterial Gene Engineering CRISPR Kit." *The Odin* website. http://www.the-odin.com/diy-crispr-kit/.

Oliver, Jack. "Earthquake Seismology in the Plate Tectonics Revolution." In *Plate Tectonics: An Insider's History of the Modern Theory of the Earth*, edited by Naomi Oreskes. London: CRC Press, 2018.

Olwell, Russell. "Physical Isolation and Marginalization in Physics: David Bohm's Cold War Exile." *Isis* 90, no. 4 (December, 1999), 745–746.

Oreskes, Naomi and Erik M. Conway. *Merchants of Doubt: How a Handful of Scientists Obscured the Truth on Issues from Tobacco Smoke to Global Warming*. New York: Bloomsbury Press, 2010.

Oreskes, Naomi. "Why I Am a Presentist." *Science in Context* 26, no. 4 (2013): 595–609.

Oreskes, Naomi. "From Continental Drift to Plate Tectonics." In *Plate Tectonics: An Insider's History of the Modern Theory of the Earth*, edited by Naomi Oreskes. London: CRC Press, 2018.

Oreskes, Naomi. *Why Trust Science?* Princeton: Princeton University Press, 2019.

Oreskes, Naomi. "Understanding the Trust (And Distrust) in Science." Interview by Ira Flatow. *Science Friday*, NPR, October 11, 2019. Audio, 24:47. https://www.sciencefriday.com/segments/naomi-oreskes-why-trust-science/.

Orwell, George. *Nineteen Eighty-Four*. New York: Harcourt, Brace and Company, 1949.

Otto, Shawn. *The War on Science: Who's Waging It, Why It Matters, What We Can Do About It*. Minneapolis, MN: Milkweed Editions, 2016.

Otto, Shawn. "A Plan to Defend Against the War on Science." *Scientific American* 9, October 9, 2016. https://www.scientificamerican.com/article/a-plan-to-defend-against-the-war-on-science/.

Outram, Dorinda. *Four Fools in the Age of Reason: Laughter, Cruelty, and Power in Early Modern Germany*. Charlottesville: University of Virginia Press, 2019.

Palmieri, Luigi. *The Eruption of Vesuvius in 1872*. London: Asher & Company, 1873.

Park, Alice. "Nobel Prize in Chemistry Awarded to First All-Female Team for CRISPR Gene Editing." *Time Magazine*, October 8, 2020. https://time.com/5897538/nobel-prize-crispr-gene-editing/.

Park, Buhm Soon. "In the 'Context of Pedagogy'." In *Pedagogy and the Practice of Science*, edited by David Kaiser, 287–319. Cambridge, MA: The MIT Press, 2005.

Peat, F. David. *Infinite Potential: The Life and Times of David Bohm*. New York: Basic Books, 1997.

Perillán, José G. "A Reexamination of Early Debates on the Interpretation of Quantum Theory: Louis de Broglie to David Bohm." PhD dissertation, University of Rochester, 2011.

Perillán, José G. "Quantum Narratives and the Power of Rhetorical Omission: An Early History of the Pilot Wave Interpretation of Quantum Theory." *Historical Studies in the Natural Sciences* 48, no. 1 (2018): 24–55.

Pietrucci, P. "Voices from the Seismic Crater in the Trial of the Major Risk Committee: A Local Counternarrative of 'the L'Aquila Seven.'" *Interface* 8, no. 2 (2016): 261–285.

Pirani, Felix A.E. "Republication of: On the Physical Significance of the Riemann Tensor." *General Relativity and Gravitation* 41 (2009): 1215–1232.

Pinch, Trevor. "What Does a Proof Do If It Does Not Prove?" In *The Social Production of Scientific Knowledge*, edited by Everett Mendelsohn et al., 171–215. Boston: D. Reidel Publishing Co., 1977.

Povoledo, E. and H. Fountain. "Italy Orders Jail Terms for Seven Who Didn't Warn of Deadly Earthquake." *The New York Times*, October 22, 2012. https://www.nytimes.com/2012/10/23/world/europe/italy-convicts-7-for-failure-to-warn-of-quake.html.

Press, William H. and Kip S. Thorne. "Gravitational-Wave Astronomy." *Annual Review of Astronomy and Astrophysics* 10, no. 1 (1972): 335–374.

Przibram, Karl, ed. *Letters on Wave Mechanics: Schrödinger, Planck, Einstein, Lorentz*. New York: Philosophical Library, 1967.

Raman, V.V. and P. Forman. "Why Was It Schrödinger Who Developed de Broglie's Ideas?" *Historical Studies in the Physical Sciences* 1 (1969): 291–314.

Rappaport, Joanne. *The Politics of Memory: Native Historical Interpretation in the Colombian Andes*. Durham: Duke University Press, 1998.

Regalado, Antonio. "Exclusive: Chinese scientists Are Creating CRISPR Babies." *MIT Technology Review*, November 25, 2018. https://www.technologyreview.com/2018/11/25/138962/exclusive-chinese-scientists-are-creating-crispr-babies/.

Rickles, D. and C.M. DeWitt, eds. *The Role of Gravitation in Physics: Report from the 1957 Chapel Hill Conference*. Communicated by Jürgen Renn,

Alexander Blum, and Peter Damerow, Edition Open Access, 2011: https://edition-open-sources.org/sources/5/.

Rickles, Dean. "The Chapel Hill Conference in Context." In *The Role of Gravitation in Physics: Report from the 1957 Chapel Hill Conference*, edited by D. Rickles and C.M. DeWitt. Communicated by Jürgen Renn, Alexander Blum, and Peter Damerow, Edition Open Access, 2011: https://edition-open-sources.org/sources/5/.

Robinson, D.C. "Gravitation and General Relativity at King's College London." *The European Physical Journal H* 44, no. 3 (2019): 181–270.

Ropeik, David. "The L'Aquila Verdict: A Judgment Not against Science, but against a Failure of Science Communication." *Scientific American 175*, October 22, 2012. https://blogs.scientificamerican.com/guest-blog/the-laquila-verdict-a-judgment-not-against-science-but-against-a-failure-of-science-communication/.

Rosen, Nathan. "On Cylindrical Gravitational Waves." In *Proceedings of the Fiftieth Anniversary Conference, Bern, July 11–16, 1955*, edited by André Mercier and Michael Kervaire. Basel: Birkhäuser, 1956.

Rosenberg, Charles. "Woods or Trees? Ideas and Actors in the History of Science." *Isis* 79, no. 4 (December 1988): 564–570.

Rosenblum, Bruce and Fred Kuttner. *Quantum Enigma: Physics Encounters Consciousness*, 2nd ed. Oxford: Oxford University Press, 2011.

Rosenfeld, Léon. Letter to David Bohm, May 30, 1952. Folder C58, David Bohm Papers, The Library, Birkbeck, University of London, UK.

Royal Swedish Academy of Sciences. "The Nobel Prize in Chemistry 2020—Genetic Scissors: A Tool for Rewriting the Code of Life." Press release, October 7, 2020. https://www.kva.se/en/pressrum/pressmeddelanden/nobelpriset-i-kemi-2020.

Russell, Sabin. "Cracking the Code: Jennifer Doudna and Her Amazing Molecular Scissors." *Cal Alumni Association, UC Berkeley*, Winter 2014. https://alumni.berkeley.edu/california-magazine/winter-2014-gender-assumptions/cracking-code-jennifer-doudna-and-her-amazing.

Sagan, Carl. *Demon-Haunted World: Science as a Candle in the Dark*. New York: Ballantine Books, 1996.

Salmon, Wesley C. *Causality and Explanation*. Oxford: Oxford University Press, 1998.

Sample, Ian. "Scientist Was Told to Remove Internet Prediction of Italy Earthquake." *The Guardian*, April 6, 2009. https://www.theguardian.com/world/2009/apr/06/italy-earthquake-predicted.

Sauer, Tilman and Raphael Scholl, eds. *The Philosophy of Historical Case Studies*. Boston Studies in the Philosophy and History of Science. Dordrecht: Springer, 2016.

Sayre, Anne. *Rosalind Franklin and DNA*. New York: Norton, 1975.

Schiff, Leonard I. *Quantum Mechanics*. New York: McGraw-Hill, 1949.

Schweber, Silvan S. "The Empiricist Temper Regnant: Theoretical Physics in the United States 1920–1950." *Historical Studies in the Physical and Biological Sciences* 17, no. 1 (1986): 55–98.

Schweber, Silvan S. "Shelter Island, Pocono, and Oldstone: The Emergence of American Quantum Electrodynamics after World War II." *Osiris* 2, no. 1 (1986): 265–302.

Schweber, Silvan. *QED and the Men Who Made It*. Princeton: Princeton University Press, 1994.

Schweninger, Lee. "Clotel and the Historicity of the Anecdote." *MELUS* 24, no. 1, African American Literature (Spring 1999): 21–36.

ScienceMag.org (AAAS). "Editors and Advisory Boards." *ScienceMag.org*. https://www.sciencemag.org/about/editors-and-editorial-boards.

ScienceMag.org (AAAS) webpage: "Science: Information for Authors," *ScienceMag.org*. https://www.sciencemag.org/authors/science-information-authors.

Servick, Kelly. "Broad Institute Takes a Hit in European CRISPR Patent Struggle," *ScienceMag*, 1, January 18, 2018. https://www.sciencemag.org/news/2018/01/broad-institute-takes-hit-european-crispr-patent-struggle.

Seth, Suman. *Crafting the Quantum: Arnold Sommerfeld and the Practice of Theory, 1890–1926*. Cambridge, MA: The MIT Press, 2010.

Seymour, Elaine and Anne-Barrie Hunter, eds. *Talking about Leaving Revisited: Persistence, Relocation, and Loss in Undergraduate STEM Education*. Switzerland: Springer, 2019.

Shapin, Steven. *The Scientific Revolution*. Chicago: The University of Chicago Press, 1996.

Shapin, Steven and Simon Schaffer. *Leviathan and the Air-Pump: Hobbes, Boyle, and the Experimental Life*, 2nd ed. Princeton: Princeton University Press, 2011.

Shapin, Steven. "Why Scientists Shouldn't Write History." Review of *To Explain the World*, by Steven Weinberg. *The Wall Street Journal*, February 13, 2015. https://www.wsj.com/articles/book-review-to-explain-the-world-by-steven-weinberg-1423863226.

Sherkow, Jacob S. "Law, History and Lessons in the CRISPR Patent Conflict." *Nature Biotechnology* 33, no. 3 (2015): 256–257.

Sherkow, Jacob S. "Patent Protection for CRISPR: An ELSI Review." *Journal of Law and the Biosciences* 4, no. 3 (2017): 565–576.

Sherkow, Jacob S. "The CRISPR Patent Landscape: Past, Present, and Future." *The CRISPR Journal* 1, no. 1 (February 1, 2018): 5–9.

Sherkow, Jacob S. "The CRISPR Patent Decision Didn't Get the Science Right. That Doesn't Mean It Was Wrong," *STAT*, September 11, 2018. https://www.statnews.com/2018/09/11/crispr-patent-decision-science/.

Silver, Nate. *The Signal and the Noise: Why So Many Predictions Fail—But Some Don't*. New York: Penguin, 2015.

Smail, Daniel Lord. *On Deep History and the Brain*. Berkeley: University of California Press, 2006.

Sommerfeld, Arnold. *Atombau und Spektrallinien, Wellenmechanischer Ergonzungsband*. Braunschweig, Vieweg, 1929.

Specter, Michael. "The Gene Hackers: A Powerful New Technology Enables Us to Manipulate Our DNA More Easily than Ever Before." *The New Yorker*, November 8, 2015. https://www.newyorker.com/magazine/2015/11/16/the-gene-hackers?irgwc=1&source=affiliate_impactpmx_12f6tote_desktop_VigLink&mbid=affiliate_impactpmx_12f6tote_desktop_VigLink.

Staley, Richard. *Einstein's Generation: The Origins of the Relativity Revolution*. Chicago: The University of Chicago Press, 2009.

Stanley, Matthew. *Practical Mystic: Religion, Science, and A.S. Eddington*. Chicago: The University of Chicago Press, 2007.

Stanley, Matthew. *Einstein's War: How Relativity Triumphed Amid the Vicious Nationalism of World War I*. New York: Dutton, 2019.

Stansfield, William D. "Luther Burbank: Honorary Member of the American Breeders' Association." *Journal of Heredity* 97, no. 2 (2006): 95–99. https://academic.oup.com/jhered/article/97/2/95/2187603.

Stefanovska, Malina. "Exemplary or Singular? The Anecdote in Historical Narrative." *SubStance* 38, no. 1 (2009): 16–30.

Stein, Rob. "New U.S. Experiments Aim to Create Gene-Edited Human Embryos." *National Public Radio*, February 1, 2019. https://www.npr.org/sections/health-shots/2019/02/01/689623550/new-u-s-experiments-aim-to-create-gene-edited-human-embryos.

Steinmetz-Jenkins, Daniel. "Beyond the Edge of History: Historians' Prohibition on 'Presentism' Crumbles Under the Weight of Events." *The Chronicle of Higher Education*, August 14, 2020. https://www.chronicle.com/article/beyond-the-end-of-history.

Stöltzner, Michael. "Opportunistic Axiomatics—Von Neumann on the Methodology of Mathematical Physics." In *John von Neumann and the Foundations of Quantum Physics*, edited by M. Rédei and M. Stöltzner. Dordrecht: Kluwer Academic Publishers, 2001.

Strevens, Michael. "The Role of the Priority Rule in Science." *The Journal of Philosophy* 100, 3 (2003): 55–79.

Tennant, Jonathan P., Jonathan M. Dugan, Daniel Graziotin, et al. "A Multi-Disciplinary Perspective on Emergent and Future Innovations in Peer Review." *F1000Research* 6 (2017). https://f1000research.com/articles/6-1151.

Thorne, Kip S. "The Generation of Gravitational Waves: A Review of Computational Techniques." In *Topics in Theoretical and Experimental Gravitation Physics*, edited by V. De Sabbata and J. Weber, 1–61. New York: Plenum Press, 1977.

Thorne, Kip S. *Black Holes and Time Warps: Einstein's Outrageous Legacy*. New York: W.W. Norton & Co., 1995.

Tosh, Nick. "Anachronism and Retrospective Explanation: In Defense of a Present-Centered History of Science." *Studies in History and Philosophy of Science Part A* 34, no. 3 (September, 2003): 647–659.

Travis, John. "Breakthrough of the Year: CRISPR Makes the Cut." *ScienceMag*, December 17, 2015. http://www.sciencemag.org/news/2015/12/and-science-s-2015-breakthrough-year.

Traweek, Sharon. *Beamtimes and Lifetimes: The World of High Energy Physicists*. Cambridge, MA: Harvard University Press, 1988.

Tresch, John. *The Romantic Machine: Utopian Science and Technology after Napoleon*. Chicago: The University of Chicago Press, 2012.

Trimble, Virginia. "Wired by Weber: the Story of the First Searcher and Searches for Gravitational Waves." *The European Physical Journal H* 42, no. 2 (2017): 261–291.

University California Berkeley, Public Affairs. "U.S. Patent Office Indicates It Will Issue Third CRISPR Patent to UC." *Berkeley News*, February 8, 2019. https://news.berkeley.edu/2019/02/08/u-s-patent-office-indicates-it-will-issue-third-crispr-patent-to-uc/.

United States Patent and Trademark Office (USPTO). "General Information Concerning Patents." *USPTO.gov*. October 2015. https://www.uspto.gov/patents-getting-started/general-information-concerning-patents.

United States Geological Survey (USGS). "Can You Predict Earthquakes?" USGS.gov. https://www.usgs.gov/faqs/can-you-predict-earthquakes?qt-_

news_science_products=0&qt-news_science_products=0#qt-news_science_products.

Vance, Tracy. "'Heroes of CRISPR' Disputed." *The Scientist: Exploring Life, Inspiring Innovation,* January 19, 2016. https://www.the-scientist.com/news-opinion/heroes-of-crispr-disputed-34188.

Vassar College Libraries. "The Albert Einstein Digital Collection at Vassar College Libraries." https://einstein.digitallibrary.vassar.edu.

Vergano, Dan. "Big Bang's 'Smoking Gun' Confirms Early Universe's Exponential Growth." *National Geographic Daily News,* March 17, 2014. http://news.nationalgeographic.com/news/2014/14/140317-big-bang-gravitational-waves-inflation-science-space/#.UymgsYXDWRg.

Vila-Valls, Adrien. "Louis de Broglie et la Diffusion de la Mécanique Quantique en France (1925–1960)." PhD dissertation, Université Claude Bernard, 2012.

Von Neumann, John. *Mathematical Foundations of Quantum Mechanics.* Princeton: Princeton University Press, 1955.

Wakefield, Andre. "Butterfield's Nightmare: The History of Science as Disney History." *History and Technology* 30, no. 3 (2014): 232–251.

Walia, Vivek, et al. "Earthquake Prediction Studies Using Radon as a Precursor in NW Himalayas, India: A Case Study." *TAO: Terrestrial, Atmospheric and Oceanic Sciences* 16, no. 4 (2005): 775–804.

Wallenborn, Grégoire and Pierre Marage, eds. *The Solvay Councils and the Birth of Modern Physics.* Basel: Birkhäuser, 1999.

Wang, Jessica. *American Science in an Age of Anxiety: Scientists, Anticommunism, and the Cold War.* Chapel Hill: The University of North Carolina Press, 1999.

Warwick, Andrew. *Masters of Theory: Cambridge and the Rise of Mathematical Physics.* Chicago: The University of Chicago Press, 2003.

Watson, James D. *The Annotated and Illustrated Double Helix.* New York: Simon & Schuster, 2012.

Weart, Spencer. *The Discovery of Global Warming: Revised and Expanded Edition.* Cambridge, MA: Harvard University Press, 2008.

Weber, Joseph and John Archibald Wheeler. "Reality of the Cylindrical Gravitational Waves of Einstein and Rosen." *Reviews of Modern Physics* 29 (1957): 509–515.

Weber, Joseph. "Detection and Generation of Gravitational Waves." *Physical Review Letters* 117 (1960): 306.

Weber, Joseph. Interview by Joan Bromberg on April 8, 1983. Niels Bohr Library & Archives, American Institute of Physics, Oral Histories,

College Park, MD. https://www.aip.org/history-programs/niels-bohr-library/oral-histories/4941.

Weinberg, Steven. *Dreams of a Final Theory*. New York: Pantheon Books, 1992.

Weinberg, Steven. "Sokal's Hoax." *The New York Review of Books*, August 8, 1996. https://www.nybooks.com/articles/1996/08/08/sokals-hoax/.

Weinberg, Steven. *To Explain the World: The Discovery of Modern Science*. London: Penguin UK, 2015.

Westfall, Richard S. *Never at Rest: A Biography of Isaac Newton*. Cambridge: Cambridge University Press, 1983.

Wheaton, Bruce. "Atomic Waves in Private Practice." In *Quantum Mechanics at the Crossroads*, edited by James Evans and Alan Thorndike. Berlin: Springer-Verlag, 2007.

White, Hayden. *Metahistory*. 40th Anniversary ed. Baltimore: Johns Hopkins University Press, 2014.

Whittaker, Andrew. "John Bell in Belfast: Early Years and Education." In *Quantum Unspeakables*, edited by R.A. Bertlman and A. Zeilinger. Berlin: Springer, 2002.

Wilson, Adrian and T.G. Ashplant. "Whig History and Present-Centered History," *The Historical Journal* 31, no. 1 (March 1988): 1–16.

Woolgar, Steve. "Writing an Intellectual History of Scientific Development: The Use of Discovery Accounts." *Social Studies of Science* 6 (1976): 395–422.

Wu, Hong-yan and Chun-yu Cao. "The Application of CRISPR-Cas9 Genome Editing Tool in Cancer Immunotherapy." *Briefings in Functional Genomics* 18, no. 2 (March 22, 2019): 129–132.

Index

Note: Figures are indicated by an italic *f* following the page number.

332 Index

Brillouin, Marcel 82
Broad Institute 110, 114, 118–21, 125, 131, 140, 146, 150
Burbank, Luther 153, 155
Bush, John W.M. 106
Butterfield, Herbert, *The Whig Interpretation of History* 25, 249n8, 255n14

Calvi, Gian Michele 225
Camilleri, Kristian 90, 271n57
Campbell, Joseph 37
Caputo, Michael 244–7
Caribou Biosciences 152
Cell 130, 133
Centro, Il 217–18, 229, 297n64
Chapel Hill "Role of Gravitation in Physics" conference (1957) 181–5
Charpak, Georges 298n76
Charpentier, Emmanuelle 121–8, 133
  Breakthrough Prize in Life Sciences 136
  Doudna and Sternberg's *A Crack in Creation* 148–50
  Kavli Prize 136–7
  Lander's "The Heroes of CRISPR" 129–32, 138, 147
  Nobel Prize 120, 126, 136, 137, 284n109
  photograph 122f
  publishing landscape 130, 133–6, 148
chimerism 11–12, 262n108
  myth-history 11–12, 58–68, 59f
China
  CRISPR-edited babies 157
  Haicheng earthquake (1975) 212
  Intellectual Property Office (SIPO) 127
Church, George 124, 125, 284n113
Cohen-Boyer patents 128, 281n51
collective memory 65–7, 68
Collins, Harry 239, 253n41
  The Golem (with Trevor Pinch) 36
  gravitational wave theory 189–93, 291nn99, 101, 292nn103, 111, 116
  history of science typology 257n28

"Two Kinds of Case Study and a New Agreement" (with Allan Franklin) 30–2, 35–6, 189
Comfort, Nathanial, "A Whig History of CRISPR" 109, 277n3
common ground 4–6, 22, 30–2, 39, 59, 70
  epistemological pluralism and standpoint theory 5, 31, 32f, 64, 78, 88, 197, 237
  example of Collins and Franklin 30-2, 189
  unpacking Einstein's black box 8–10, 9f, 22, 29, 46, 244
consensus 12, 18–20, 31, 56, 67–9, 78–80, 90–1, 153, 181, 189, 206–7, 232, 236–7
continental drift theory 211
Conway, Erik M. 254n52
  Merchants of Doubt (with Naomi Oreskes) 232, 302n15
Copernicus 44
Córdova, France 161–3, 166, 168–9, 195–6, 293n121
  photographs 124f, 162f
cosmic microwave background (CMB) 238–9
cosmologies 198
COVID-19 pandemic 6–7, 243–7
Crew, Henry 41–2
Crick, Francis 142–4
CRISPR 15–16, 157, 242
  basics 110–13
  Doudna and Sternberg's *A Crack in Creation* 139–42, 145–55, 157, 282n74
  Lander's "The Heroes of CRISPR" 107–10, 120–1, 124, 128–9, 136, 152, 154, 157, 281n53
  downplaying of Doudna and Charpentier 129–32, 138, 151
  lessons learned 137–9
  map 116–20, 117f, 129, 150
  miraculous scientific ecosystem 113–16, 116f
  patents 120–9
  publishing landscapes 132–7, 148
  schematic illustration 112f

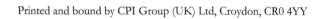
Printed and bound by CPI Group (UK) Ltd, Croydon, CR0 4YY